I0523920

William Hurrell Mallock

Aristocracy and Evolution

A Study of the Rights, the Origin, and the Social Functions of the Wealthier Classes

William Hurrell Mallock

Aristocracy and Evolution
A Study of the Rights, the Origin, and the Social Functions of the Wealthier Classes

ISBN/EAN: 9783337005191

Printed in Europe, USA, Canada, Australia, Japan

Cover: Foto ©Suzi / pixelio.de

More available books at **www.hansebooks.com**

ARISTOCRACY & EVOLUTION

A STUDY OF THE RIGHTS, THE ORIGIN, AND THE SOCIAL FUNCTIONS OF THE WEALTHIER CLASSES

BY

W. H. MALLOCK

AUTHOR OF 'IS LIFE WORTH LIVING?' 'A HUMAN DOCUMENT,'
'LABOUR AND THE POPULAR WELFARE,' ETC.

Toute civilisation est l'œuvre des aristocrates.
RENAN.

'Tis thus the spirit of a single mind
  Makes that of multitudes take one direction,
As roll the waters to the breathing wind,
  Or roams the herd beneath the bull's protection,
Or as a little dog will lead the blind,
  Or a bell-wether form the flock's connection
By tinkling sounds, when they go forth to victual,
Such is the sway of your great men o'er little.

There was not now a luggage-boy but sought
  Danger and spoil with ardour much increased ;
And why ? Because a little—odd—old man,
Stript to his shirt, was come to lead the van.
BYRON.

LONDON
ADAM AND CHARLES BLACK
1898

PREFACE

THE word *aristocracy* as used in the title of this volume has no exclusive, and indeed no special reference to a class distinguished by hereditary political privileges, by titles, or by heraldic pedigree. It here means the exceptionally gifted and efficient minority, no matter what the position in which its members may have been born, or what the sphere of social progress in which their exceptional efficiency shows itself. I have chosen the word *aristocracy* in preference to the word *oligarchy* because it means not only the rule of the few, but of the best or the most efficient of the few.

Of the various questions involved in the general argument of the work, many would, if they were to be examined exhaustively, demand entire treatises to themselves rather than chapters. This is specially true of such questions as the nature of men's congenital inequalities, the effects of different classes of motive in producing different classes of action, and the effects of equal education on unequal talents and temperaments. But the practical bearings of an argument are more readily grasped when its various parts are set forth with comparative brevity, than they are when the attention claimed for each is minute enough to do it justice as a separate subject of inquiry; and it has appeared to me that in the present condition of opinion, prevalent social fallacies may be more easily combated by putting the case against them in a form which will render it intelligible to everybody, and by leaving many points to be elaborated, if necessary, elsewhere.

I may also add that the conclusions here arrived at, with whatever completeness they might have been explained, elaborated, and defended, would not, in my opinion, do more than partially answer the questions to which they refer. This volume aims only at establishing what are the social rights and social functions, in progressive communities, of the few. The entire question of their duties and proper liabilities, whether imposed on them by themselves or by the State, has been left untouched. This side of the question I hope to deal with hereafter. It is enough to observe here that it is impossible to define the duties of the few, of the rich, of the powerful, of the highly gifted, and to secure that these duties shall be performed by them, unless we first understand the extent of the functions which they inevitably perform, and admit frankly the indefeasible character of their rights.

CONTENTS

BOOK I

CHAPTER I

THE FUNDAMENTAL ERROR IN MODERN SOCIOLOGICAL STUDY

CHAPTER II

THE ATTEMPT TO MERGE THE GREAT MAN IN
THE AGGREGATE

CHAPTER III

Great Men, as the true Cause of Progress

CHAPTER IV

THE GREAT MAN AS DISTINGUISHED FROM THE PHYSIOLOGICALLY FITTEST SURVIVOR

BOOK II

CHAPTER I

THE NATURE AND THE DEGREES OF THE SUPERIORITIES OF GREAT MEN

CHAPTER II

PROGRESS THE RESULT OF A STRUGGLE NOT FOR SURVIVAL, BUT FOR DOMINATION

CONTENTS xvii

CHAPTER IV

THE MEANS BY WHICH THE GREAT MAN ACQUIRES POWER IN POLITICS

b

BOOK III

CHAPTER I

How to discriminate between the Parts contributed to a Joint Product by the Few and by the Many

CHAPTER II

THE NATURE AND SCOPE OF PURELY DEMOCRATIC ACTION, OR THE ACTION OF AVERAGE MEN IN CO-OPERATION

CONTENTS xxiii

CHAPTER III

THE QUALITIES OF THE ORDINARY AS OPPOSED TO THE GREAT MAN

BOOK IV

CHAPTER I

THE DEPENDENCE OF EXCEPTIONAL ACTION ON THE ATTAIN-ABILITY OF EXCEPTIONAL REWARD, OR THE NECESSARY CORRESPONDENCE BETWEEN THE MOTIVES TO ACTION AND ITS RESULTS.

CHAPTER II

The Motives of the Exceptional Wealth-Producer

CHAPTER III

EQUALITY OF EDUCATIONAL OPPORTUNITY

CHAPTER IV

INEQUALITY, HAPPINESS, AND PROGRESS

BOOK I

CHAPTER I

THE FUNDAMENTAL ERROR IN MODERN
SOCIOLOGICAL STUDY

THE interest with which the world in general, Science during the middle of throughout the middle portion of this century, this century has watched the progress of the various positive excited popular interest mainly sciences, would, when we consider how abstruse on account of its bearing on these sciences are, seem strange and almost in- the doctrines of Christianity. explicable if it were not for one fact. This fact is the close and obvious bearing which the conclusions of the sciences in question have on traditional Christianity, and, indeed, on any belief in immortality and the divine government of the world. The popular interest in science remains still unabated, but the most careless observer can hardly fail to perceive that the grounds of it are, to a certain extent, very rapidly changing. They are Its popularity now is beginceasing to be primarily religious, and are becoming ning to depend on its bearing primarily social. The theories and discoveries of not on religious the *savant* which are examined with the greatest problems, but on social. eagerness are no longer those which affect our prospects of a life in heaven, but those which deal with the possibility of improving our social conditions on earth, and which appeal to us through our sympathies, not with belief or doubt, but with the principles which are broadly contrasted under the names of conservative and revolutionary.

Science itself is undergoing a corresponding change.

Such being the case, it is hardly necessary to observe that science itself has been undergoing a change likewise. The character of the change, however, requires to be briefly specified. From the time when geologists first startled the orthodox by demonstrating that the universe was more than six thousand years old, and that something more than a week had been occupied in the process of its construction, to the time, comparatively recent, during which the genius of Darwin and others was forcing on the world entirely new ideas with regard to the parentage, and presumably the nature of man, there was a certain limit—a certain scientific frontier — at which positive science practically stopped short. Having sedulously examined the

Its characteristic aim during the middle of the century was to deal with physical and physiological evolution.

materials and structure of the universe, until on the one hand it reached atoms and molecules, it examined, on the other, the first emergence of organic life, and traced its developments till they culminated in the articulate-speaking human being. It brought us, in fact, to man on the threshold of his subsequent history; and there, till very recently, positive science left him. But now there are signs all round us of a new intellectual movement, analogous to that which accompanied the rise of Darwinism,

and science once again is endeavouring to enlarge
its borders. Having offered us an explanation of
the origin of the animal *man*, it proposes to deal
with the existing conditions of society very much as
it dealt with the structure of the human body, to
exhibit them as the necessary result of certain far-
reaching laws and causes, and to deduce our
civilisation of to-day from the condition of the
primitive savage by the same methods and by the
aid of the same theories as those which it employed
in deducing the primitive savage from the brutes,
and the brutes in their turn from primitive germ
or protoplasm. In other words, the great triumph
of science during what we may call its physical Its characteristic aim now
is to deal with
the evolution
of society.
period has been the establishment of that theory
of development which is commonly spoken of as
Evolution, and the application of this to the problems
of physics and biology. The object of science in
entering on what we may call its social period is the
application of this same theory to the problems of
civilisation and society.

It is true that, if we use the word science in a
certain sense, the attempt to treat social problems
scientifically is not in itself new. Political economy,
to say nothing of utilitarian ethics, is a social science,
or it is nothing ; and political economy had already
made considerable advances when modern physical
science had hardly found its footing. But before
long physical science passed it, with a step that was
not only more rapid, but also immeasurably firmer,
and was presently giving such an example of what

accurate science is, that it was thought doubtful whether political economy could be called a science at all. The doubt thus raised cannot be said to have justified itself. In spite of all the attacks that have been made against the earlier economists, their principal doctrines survive to the present day, as being, so far as they go, genuine scientific truths. But whenever the thinker, who has been educated in the school of modern physical science, betakes himself now to the study of society and human action, and begins to apply to these the developed theory of evolution, though he does not reject the doctrines of the earlier economists, he sees them in a new light, by which their significance is profoundly changed. The earlier economists took society as they found it, and they reasoned as though what was true of the economic life around them must be absolutely and universally true of economic life always. Here is the point as to which the thinker

What is new is
the application
to it of the
evolutionary
theory.

of to-day differs from them. He does not dispute the truth of the deductions drawn by them with regard to society as it existed during their own epoch ; but, educated by the methods and discoveries of the physical and biological evolutionist, he perceives that society itself is in process of constant change, that many economic doctrines which have been true during the present century had little application to society during the Middle Ages, and that centuries hence they may perhaps have even less. Thus, though he does not repudiate or disregard the economic science of the past, he merges it in a science the scope of which is far wider and deeper. This is a science which primarily sets itself to explain, not how a given set of social conditions affects those who live among them, but how social conditions at one epoch are different from those of another, how each set of conditions is the resultant of those preceding it, and how, since the society of the present differs from that of the past, the society of the future is likely to differ from that of the present.

What political economy has thus lost in precision it has gained in general interest. So long as it merely analysed processes of production and distribution which it was assumed would always continue without substantial modification, political economy was mainly a science for specialists, and was little calculated to arouse any keen interest in the public. But now that it has been merged in that general science of evolution, which offers to an unquiet age what seems a scientific licence to regard as practically producible some indeterminate transformation in these processes, political economy has come to occupy a new position. Instead of being ignored or ridiculed by the more ardent school of reformers, and even neglected by conservatives as a not very powerful auxiliary, it has now been brought down into the dust of the general struggle, and is invoked by one side as the prophetess of new possibilities, and by the other as an exorcist of mischievous and mad illusions. And what is true in this respect with regard to political economy is

Book I
Chapter I

This excites men by suggesting great social changes in the future,

Book 1
Chapter 1

also true with regard to evolutionary social science as a whole. Social science as a whole, just like this special branch of it, is being brought into vital contact with the lives and hopes of man, and is exciting a popular interest strictly analogous to that which had been excited by physical and biological science previously.

which will give a religious meaning to the history of humanity,

It · is doing this in two ways, which, though closely connected, are distinct. In the first place, it is directing our attention to the human race as a whole, and is showing us how society and the individual have developed in an orderly manner, growing upwards from the lowest and the most miserable beginnings to the heights of civilisation, intellectual, moral, and material, and how they contain in themselves the potency of yet further development. It thus offers to the mind a vast variety of suggestion with regard to the significance of man's presence upon the earth, and is held by many to be supplying us with the materials of a religion calculated to replace that which physical science has discredited. The second way in which it excites

or secure for men now existing, or for their children, practical social advantages.

popular interest is the way which has been just illustrated by a reference to political economy. For besides offering to our philosophic and religious faculties the vision of man's corporate movement from a condition of helpless bestiality towards some "far-off divine event," which glitters on us in the remote future, social science is suggesting to us changes which are of a very much nearer kind, and which appeal not to our speculative desire to discover some meaning in the universe, but to the personal interest which we each of us take in our own welfare—such, for instance, as a general redistribution of wealth, the abolition or complete reorganisation of private property, the emancipation of labour, and the realisation of social equality.

*Book I
Chapter 1*

This distinction between the speculative and practical aspects of social science has a special importance, which will be explained and insisted on presently. But it is here mentioned only to show the reader how strong a combination of motives is impelling the present generation—the conservative classes and the revolutionary classes equally—to transfer to social science the interest once felt in physical; and how strong is the stimulus thus applied to sociologists to emulate the diligence and success of the physicists and biologists, their predecessors.   Nor have diligence, enthusiasm, or scientific genius been wanting to them.   As has already been observed, they have transformed social science altogether by applying to it the doctrines of evolution which physical science taught them, and have thus organically affiliated the former study to the latter.   This is in itself a triumph worthy of the enterprise that has achieved it.   But they have done far more than borrow from physics this mere general theory.   They have established between physical phenomena and social an enormous number of analogies, so close that the one set assists in the interpretation of the other.   They have borrowed from the physicists a number of their subsidiary theories, their methods of grouping facts, and, above

Men have thus a double reason for being interested in social science, and sociologists a double reason for studying it;

and it has attracted a number of men of genius, who have applied to it the methods learned in the school of physical science.

all, their methods of studying them. In a word, they are endeavouring to follow the masters of physical science along the precise path which has led the latter to such solid and such definite results.

Yet despite their genius and their diligence, all parties complain that the results of their study are inconclusive.
We have now, however, to record a singular and disappointing truth. Though men of science have, in the manner just described, been engaged for years in the field of sociological study; though the way was prepared for them by men like Comte, Mill, and Buckle; though amongst them have been men like Mr. Spencer, with capacities of the highest order, and though certain results have been reached of the kind desired, complaints are heard from thinkers of all shades of opinion that these results are singularly unsatisfactory and inconclusive when compared with the efforts that have been made in reaching them, and still more when compared with the results of corresponding efforts in the sphere of physics.

Professor Marshall and Mr. Kidd, for instance, complain of the fact, but can suggest no explanation of it.
No one complains more loudly of this comparative failure than some of the most distinguished students of social science themselves. Professor Marshall, for instance, who has done more than any other English author to breathe into technical economics the spirit of evolutionary science, admits that Comte, who laid the foundation of sociology, and Mr. Spencer, who has invested it with a definitely scientific character, have brought to the study of *"man's actions in society unsurpassed knowledge and great genius, and have made epochs in thought by their broad surveys and suggestive hints"*; but neither of them, he proceeds to say, has succeeded

in doing more than this. Mr. Kidd, again, whose
work on *Social Evolution*, if not valuable for the
conclusions he himself desires to substantiate, is
curiously significant as an example of contemporary
sociological reasoning, repeats Professor Marshall's
complaint, and gives yet more definite point to it.
Having observed that "*despite the great advance
which science has made in almost every other direction, there is, it must be confessed, no science of human
society, properly so called,*" he justifies this observation by insisting on what is an undoubted fact, that
"*so little practical light has even Mr. Herbert Spencer
succeeded in throwing on the nature of the social
problems of our time, that his investigations and
conclusions are, according as they are dealt with by
one side or the other, held to lead up to the opinions
of the two diametrically opposite camps of individualists and collectivists, into which society is rapidly
becoming organised.*"

Now what is the reason of this? Here is the
question that confronts us. That the methods
adopted by the scientist in the domain of physics
are applicable to social phenomena, just as they are
to physical, has been not only established in a
broad and general way, but demonstrated by a mass
of minute and elaborately co-ordinated facts. Why,
then, when we find them in the sphere of physics
solving one problem after another with a truly
surprising accuracy, do they yield us such vague
and often contradictory results when we apply them
to the solution of the practical problems of society?

Book I
Chapter 1

The answer will be found in the fact just referred to— that social science attempts to answer two distinct sets of questions ;

Those who complain so justly of the failure of social science and who yet show themselves altogether at a loss to account for it, might have seen their way to answering this question had they concentrated their attention on a point that was just now alluded to. It was just now observed that the problems which social science aims at answering, and is popularly expected to answer, are of two distinct kinds—the philosophic or religious, and the practical ; the former being concerned with the destinies of humanity as a whole, with movements extending over enormous periods of time, and with the remote past and future far more than with the present ; the other being concerned exclusively with the present or the near future, and with changes that will affect either ourselves or our own children.

and one set— namely the speculative— it has answered with great success ;

Now it will be found that social science, whilst busying itself with both these sets of problems, has met with the failures which are alleged against it, only in dealing with the latter, and that, so far as regards the former, it has successfully reached conclusions comparable in precision and solidity to those of the physicists and biologists whose methods it has so conscientiously followed. Professor Marshall's own treatise on *The Principles of Economics*, and that of Mr. Kidd on *Social Evolution* likewise, abound in admissions that this statement of the case is correct. Professor Marshall's account of the rise and fall of civilisation as caused by climate, by geographical position, and the influence of one race and one civilisation on another,—an account of which he

places in the very forefront of his elaborate work
—is professedly merely a summary of conclusions
already arrived at ; and the manner in which he
states these conclusions is itself evidence that
sociologists, when dealing with certain classes of
social phenomena, have given us something more
than " *surveys* " and " *suggestive hints.*" Social
science, in fact, cannot be properly called a failure
except when it ceases to deal with the larger it has failed
phenomena of society, which show themselves only only in attempting to
in the long course of ages, and descending to the answer practical questions.
problems of a particular age and civilisation, endeavours to deduce, from the general principles it
has established, propositions minute enough to be
applicable to our immediate conduct and expectations. As practical inquirers, therefore, the real
question before us is not why social science has
failed, where physical science has succeeded, but
why social science has succeeded like physical science
in one direction, and, unlike physical science, failed
so signally in another. If we concentrate our
attention on the subject in this way, and thus
realise with precision the nature of the failure we
desire to explain, we shall find that the explanation
of it is not only far simpler than might have been
supposed, but also that the remedy for it is far more
obvious and more easy.

It has been said that sociology has succeeded in Now the
phenomena
dealing with those social phenomena which extend with which it
has dealt with
themselves through vast periods of time, and has successfully
failed in dealing with those whose interest and are phenomena of social

existence is limited to lives of a few particular generations. Now between these two sets of phenomena, as thus far described, the most obvious difference is, no doubt, the difference in their magnitude. This difference, however, is altogether accidental, and does nothing to explain those curiously contrasted results which the study of one set and the other has yielded to the modern sociologist. The difference, which will explain these, is of quite another kind, and may briefly be stated thus. The larger social phenomena—those which interest the speculative philosopher, and with which sociology has dealt successfully, are phenomena of social aggregates, or masses of men regarded as single bodies ; the smaller phenomena—those which interest the practical man, and with which sociology has dealt unsuccessfully—are essentially the phenomena not of social aggregates, but of various parts of aggregates.

Let us illustrate the matter provisionally by two rudimentary examples. As an example of the larger phenomena let us take the advance of man from the age of stone to the ages of bronze and iron. Of the smaller, we may take the phenomena referred to by Mr. Kidd—namely, the appearance in the modern world of the socialist or collectivist party, and the antagonism between it and the party of private property and individualism. Now the first of these two sets of phenomena—the use by men of stone implements, and the subsequent use of metal implements —consist of phenomena which, so far as the socio-

logist is concerned, are manifested successively by
humanity, or some portion of humanity, as a whole.
They are not referred to individuals or small classes.
No question is asked as to what particular savage
may rightly claim priority in the invention of metal
implements, or whether flint or bronze were the
subjects of any prehistoric monopoly. Those races
amongst which the use of the metals became general
are regarded as a single body, which had made this
advance collectively. They are, indeed, as we shall
again have occasion to observe, habitually described
under the common name of *Man*. But let us turn But the practi-
to such phenomena as the antagonism between in- cal problems
of to-day, with
dividualists and collectivists, and the case is wholly which it has
dealt unsuc-
different. It is true that here also, as in the case cessfully, arise
out of the con-
we have just been considering, our attention is flict between
different parts
called to a portion of the human race, namely, the of aggregates.
Western or progressive nations, which we may, for
certain purposes, regard as a single aggregate ; but
it is fixed, not on the phenomena which this aggregate exhibits as a whole, but on those exhibited
by unlike and conflicting parts of it—the part which
sympathises with individualists on the one hand,
and the part which sympathises with collectivists
on the other.

Thus the subject-matter of sociology, regarded
as a speculative science, consists of those points
in which the members of any given social aggregate resemble one another. The subject - matter
of sociology, regarded as a practical science,
consists of those points in which the members, or

certain groups of members, of any given social aggregate differ from one another. And here we come to the reason why sociology, as a practical

Social science
has failed as a
practical guide
because it has
not recognised
this distinc-
tion ;
science, has failed. It has failed because hitherto it has not realised this distinction, and has persisted in applying to the phenomena, involved in practical social problems, the same terminology, the same methods of observation and reasoning, which it has applied to the phenomena involved in speculative social problems. By so doing, though it has dis-sipated many popular errors, it has, in the most singular manner, given a new vitality to others. It has indeed supplied a pseudo-scientific sanction to the most abject fallacies that have vitiated the political philosophy of this century ; and it has thus been

and hence
arise most of
the errors of
the political
philosophy of
this century.
instrumental in keeping alive and encouraging the most grotesquely impossible hopes as to what may be accomplished by legislation, and the most grotesquely false views as to the sources of social and political power. To expose these fallacies, and the defective reasoning on which they rest, is the object of the present volume.

The nature of that peculiarity in the procedure of modern sociology which has just been described, and to which all its errors are due, forms a very curious study, and it will be essential to exhibit it with the utmost plainness possible. In the following chapter, therefore, the reader shall be presented with examples of it.

CHAPTER II

THE ATTEMPT TO MERGE THE GREAT MAN IN
THE AGGREGATE

LET us take any book we please, by any modern writer, who is attempting to deal with any social subject scientifically, and whenever he is calling attention to the great intellectual triumphs which have caused the progress of civilisation, or to any developments of human nature which have marked it, we shall find that these triumphs or developments are always attributed indiscriminately to the largest mass of people with whom they have any connection —sometimes to "the nation," sometimes to "the age," sometimes to "the race," and more frequently still to "man." *Whatever may be done by some men, or classes of men, sociologists are at present accustomed to attribute to man.*

Reference has been made already to Mr. Kidd's work on *Social Evolution*, which, on its publication, attained an extraordinary popularity, and which, whatever its value otherwise, is interesting as a type of contemporary sociological reasoning. It is peculiarly interesting as illustrating the point which we are now discussing. Most of Mr. Kidd's reasoning, especially in the crucial parts of it, is not *Mr. Kidd's Social Evolution, for instance, is based entirely on this procedure.*

only conducted, but is actually represented by a
terminology which refers everything to "the race,"
"the age," or "man."   And it would be hard to
find better examples in the works of any other
writer of the condition of thought underlying the
use of these phrases, and of the extraordinary
consequences to which it leads.

He quotes with
approval two
other writers
who have been
guilty of it, Three examples will be enough.   The two first
shall be from two other writers, whom Mr. Kidd
quotes with admiration; the third shall be from
himself.   We will begin with the following passage,
taken from a contemporary economist, which Mr.
Kidd singles out for emphatic approval as "*a very
effective statement*" of one of the truths of social
science.

"*Man*," so the passage runs, "*is the only animal
whose wants can never be satisfied.   The wants of
every other living thing are uniform and fixed.
The ox of to-day aspires no more than did the ox
when man first yoked him. . . . But not so with
man* [*himself*].   *No sooner are his animal wants
satisfied, than new wants arise. . . .* [*He*] *has but set
his feet on the first step of an infinite progression.
. . . It is not merely his hunger, but taste, that
seeks gratification in food. . . . Lucullus will sup
with Lucullus; twelve boars roast on spits that
Antony's mouthful of meat may be done to a turn;
every kingdom is ransacked to add to Cleopatra's
charms; and marble colonnades, and hanging
gardens, and pyramids that rival the hills, arise.*"
This passage is taken from Mr. Henry George.

Our second example shall be a passage which Mr.
Kidd has borrowed from a far more educated thinker—M. Emile de Lavelaye. Mr. Kidd quotes M. de Lavelaye as saying that the eighteenth century brought the following message to "*man.*" "*Thou shalt cease to be the slave of the nobles and despots who oppress thee. Thou shalt be free and sovereign.*" But the realisation of the promise thus given has, in the present century, he goes on to say, confronted us with this strange problem, "*How is it that the Sovereign often starves? How is it that those who are held to be the source of power often cannot, even by hard work, provide themselves with the necessaries of life?*"

Now all these passages, if we consider them carefully, will be seen to consist of statements, every one of which is false to fact. To say that man's wants are less stationary than those of the ox is not even rhetorically true, unless we mean by "*man*" certain special races of men ; whilst the statements that follow are not true, rhetorically or otherwise, of any race at all, but only of scattered individuals. A really fine and discriminating taste in food is, as every epicure knows, rare even amongst the luxurious classes. Antony and Lucullus are types of what is not the rule, but the exception. So too are the individuals who either desire hanging gardens, or could design them ; and more exceptional still are the individuals whose personal pride and power either desire or can secure the erection of pyramids for their tombs.

Book I
Chapter 2

and the con-
sequences to
their reasoning
are ludicrous.
In M. de Lavelaye's utterances there is an analogous misstatement and misconception of every fact with which he deals. The promises of political democracy, as he describes them, were never addressed to "*man*," nor ever professed to be. The whole point of them was that they were addressed to certain classes of men only ; and that, as addressed to other classes, they were not promises, but threats. But a still graver confusion arises when the "*Sovereign*" is spoken of as starving. If by the "*Sovereign*" M. de Lavelaye really means "*Man*" as a whole, it is perfectly obvious that the "*Sovereign*" never starves. The statement is equally untrue if the Sovereign is taken to mean not man as a whole, but the immense majority of men ; and to ask why the Sovereign often does something which it never does, is not to formulate an actual problem loosely, but to convert an actual problem into one that is quite imaginary. The actual problem is not why the whole or the immense majority of mankind often starves, but why there are nearly always small sections of men who do so, the majority all the while obtaining its normal nutriment ; and the absurd result of confusing these two very different things is seen in the second form which M. de Lavelaye gives his question. "*How is it*," he asks, "*that those who are held to be the source of power often cannot, even by hard work, provide themselves with the necessaries of life ?*" The answer is that the particular groups of workers who, at any given time, happen to be unemployed,

were never held to be *the* source of power by anybody. M. de Lavelaye might as well take one
half of the passengers on a Dover packet, and
treating them as identical with the British nation
at large, ask how it is that those who are held to
rule the waves can hardly set foot on a deck without
clamouring for the steward's basin.

And now let us turn to Mr. Kidd himself. The
object of his book is to vindicate supernatural
religion by exhibiting it as advantageous to its
possessors in the social struggle for existence. He
endeavours to make good his position by two distinct
lines of argument. The first of these is that the
social struggle for existence, though it produces
progressive communities, and communities fitted to
endure, is injurious to the majority of those who at
any given time are engaged in it, and benefits only
a minority, described by him as "*the power-holding
classes.*" This minority, according to his account,
could always, if it pleased, as it has pleased in all
former ages, defend its position and keep the
majority in subjection; but it is now beginning,
under the pressure of a religious impulse, to
surrender to its inferiors voluntarily advantages
which they could never have extorted from it; and
in this great fact our hope for the future lies.

Such is one of the two main portions of Mr.
Kidd's message to the world; and here follows the
other, which will be found to be fundamentally
inconsistent with it. "*Man,*" if he had chosen to
do so, Mr. Kidd maintains—and this assertion

Mr. Kidd's reasoning itself is not less ludicrous. The first half of his argument is that religion prompts the few to surrender advantages to the many, which, if they chose to do so, they could keep.

The second half is that the many at any time could have taken these advantages from the few, and that religion alone prevented them from doing so.

is repeated by him with the utmost precision and emphasis—could at any period in his history have "*suspended the struggle for existence*" and "*organised society on a socialistic basis*"; and seeing that the struggle for existence, although essential to progress in the long-run, is injurious to the majority of each generation that takes part in it, man, if his chief guide had been reason or self-interest, would have been suspending this struggle constantly for the sake of his own present advantage, and leaving the future to take care of itself. Now, seeing that he does not, as a fact, pursue this obviously reasonable course, it follows that some power opposed to reason must have withheld him; and this power, argues Mr. Kidd, can be nothing else than religion. Here, he says, are the two functions of religion in evolution. It induces man to submit to the hardships of the evolutionary struggle, at the same time it redeems him from them by softening the hearts of the minority.

This contradiction is entirely due to the fact that, having first divided the social aggregate into two classes, he then obliterates his division, and thinks of them both as "*man*."

Now with Mr. Kidd's views about religion we have nothing to do here. We are concerned only with the extraordinary self-contradiction involved in these his principal lines of argument, and also with the cause which has led to it, and made it possible. At one moment he says that the majority in all progressive communities have been forced to submit to conditions of life that are prejudicial to them, by a powerful minority to whom these conditions are beneficial, and who, if they chose to do so, would still be able to maintain them. At

another moment he says that this surprisingly patient
majority could have easily "*suspended these conditions*" at any period of its history, and only failed
to do so because religion prompted it to forbear.
How a contradiction of this kind could have found
its way into the reasoning of a really painstaking
thinker, and been actually allowed to form the back-
bone of it, may at first sight seem inexplicable ; but
it is simply a typical result of the practice we are
now considering—that practice, common to all our
modern sociologists, of grouping the men they deal
with into the largest aggregate possible, and treating
mixed classes of men as one single class—"*man.*"

It is easy to see precisely how Mr. Kidd's mind
has worked.   In the first part of his argument he
divides progressive communities into two sections,
which he calls respectively "*the power - holding
classes*" or the "*successfuls,*" and the "*excluded
classes*" or the "*unsuccessfuls*"; and he declares
that the latter would naturally desire to suspend
the conditions of progress, whilst the former would
naturally desire, and are also able to maintain them.
But when he pushes his argument farther, and
advances to the proposition that if reason had been
"*man's*" sole guide, the conditions of progress would
have been suspended over and over again, he is
enabled to take this extraordinary step only because
his thought and his terminology undergo an unconscious metamorphosis.   He forgets his original
analysis altogether.   He merges the two classes, so
sharply contrasted by him, into one.   He argues and

thinks about them both, under the single category of "*man*"; he builds up his conclusions by joining together the very things which, in arranging his premises, he had so carefully put asunder ; and the ·result of his speculation reduced to its simplest terms is this—that "*man*" could have done, at any period of his history, and if reason had been his sole guide, actually would have done, something that was against the interests of the stronger part of him, and beyond the power of the weaker.

Mr. Kidd's
confusion is
the result of
no accidental
error. It is
the inevitable
result of a
radically
fallacious
method,

The reader will not find much difficulty in understanding that if sociologists persist in reasoning thus, they are hardly likely to arrive at any conclusion sufficiently definite to guide us in the practical difficulties of life. It may be urged, however, that such language as we have been considering, though used by scientific writers, is intended itself to be rhetorical rather than scientific, or that it betrays the inaccuracy of this or that individual thinker, instead of arising from a fundamental error in method. If any one thinks this, he shall soon be disabused of his opinion. The reader shall now be presented with a brief summary of the method deliberately followed, and of some of the conclusions arrived at by that distinguished

thinker who has done more than any one else to impart to sociology the character which it at present possesses ; and the error which lies at the bottom of the reasoning we have been just considering shall there be exhibited, systematically exemplified, and explicitly and elaborately defended. It is perhaps

hardly necessary to say that the thinker thus referred
to is Mr. Herbert Spencer.

We will then follow Mr. Spencer's reasoning
from the beginning, as set forth in his works ; and
before consulting his monumental *Principles of*
Sociology we will turn to his *Study of Sociology*, a
smaller and preparatory treatise, in which the
methods adopted by him in his main inquiry are
explained.   He opens this treatise with declaring
that until recent years any scientific treatment of
social phenomena was impossible ; and it was impossible, he says, for two definite reasons.   These
were the prevalence of two utterly false theories,
both of which precluded the idea that anything like
law or order of a calculable kind were prevalent in
the social sphere.   One of these theories was "*the
theocratic theory*," the other what he calls "*the
great-man theory*."

The theocratic theory is that which explains all
social change by reference to the direct and arbitrary
interference of a Deity ; and if this be adopted,˙ Mr.
Spencer has no difficulty in showing that anything
like a social science must be necessarily looked on
as impossible : for the only thread by which social
phenomena are connected will in that case be
hidden in the will of an inscrutable Being, which
may indeed be made known to us by revelation, but
which is not susceptible of being either observed or
calculated.   This theory, however, in its cruder
form, at all events, is, says Mr. Spencer, being fast
discarded by everybody—even by the theologically

orthodox; and the really important foe which social science has to fight against is the great-man theory, not the theocratic. Accordingly, it is by a criticism of the great-man theory that he introduces us to the theory of society, which is in his estimation true, and which alone presents social phenomena to us as amenable to scientific treatment.

The great-man theory is summed up by him in the following quotation from Carlyle: "*As I take it, universal history, the history of what man has accomplished in this world, is at bottom the history of the great men who have worked here.*" "*This,*" observes Mr. Spencer, "*not perhaps distinctly formulated, but everywhere implied, is the belief in which nearly all are brought up*"; and it is, he declares, as incompatible as the theocratic theory itself with any belief in the possibility of a social science, or any comprehension of what such a science is; for either the great man is regarded as the miraculous instrument of the Deity, a kind of "*deputy - God,*" in which case we have "*theocracy once removed*"; or else his greatness, though regarded as a natural phenomenon, is regarded as one whose occurrence is so far fortuitous, that a great man of any given kind of greatness might appear in one age or nation just as well as in another; and in this case, if social changes depend on the great man's actions, these changes will be as fortuitous as the great man's own appearance, and will as little admit of any scientific calculation.

If, however, the great man is regarded as a

for if the appearance of the great man is incalculable, progress, if it depends on him, must be incalculable also;

natural phenomenon at all, if he is not to be looked
upon as a species of incalculable angel, this idea
of his fortuitous appearance is, says Mr. Spencer, but if the great
plainly quite untenable.   The great man, unless he man is not a
miraculous
differs miraculously from other men, is produced as apparition, he
owes his great-
they are, in accordance with natural laws, and, like ness to causes
outside him-
them, owes his greatness to his near and remote self;
progenitors, just as a negro owes to his, his facial
angle, his blackness, and his woolly hair. " *Who
would expect*," Mr. Spencer asks, "*that a Newton
might be born of a Hottentot family, or that a
Milton might spring up among the Andamanese ?* "
The theory, then, which explains social changes by
referring them to the great men whose names are
connected with their initiation, will, unless it is
regarded as a theory of perpetual miracle, be
recognised as inadequate, even by those who have
hitherto held it, when once they have realised the
absurd supposition which it implies.   The great
man, whatever his seeming influence, is merely the
agent of other influences which are behind him.
He merely transmits a shock, like a man pushed
by a crowd.   Even supposing what Mr. Spencer
entirely denies to be the case, that he could really and it is these
causes which
" *remake his society*," his society none the less must really produce
have previously made him, and supplied him with the effects of
which he is the
those conditions which rendered his career possible ; proximate
initiator.
and therefore, of any changes which he may popularly be said to have caused, he is merely " *the
proximate initiator*," not the true cause at all ; and
"*if*," says Mr. Spencer, "*there is to be anything*

like a real explanation of such changes, it must be sought (not in the great man himself), but in the aggregate of social conditions, out of which he and they have arisen." Except, perhaps, in the military struggles of primitive savage tribes, "*new institutions, new activities, new ideas, all,*" he says,

"*unobtrusively make their appearance, without the aid of any king or legislator; and if you wish to understand the phenomena of social evolution, you will not do it, should you read yourself blind over the biographies of all the great rulers on record, down to Frederick the Greedy, and Napoleon the Treacherous.*" And he points his moral by observing, with a certain philosophic tartness, that there is no surer index of a man's "*mental sanity*" than the degree of contempt which, as a scientific thinker, he feels for the class of facts which the biography of individuals offers him.

Such, then, being Mr. Spencer's theory of the way in which social phenomena must be regarded, if we mean to make them the subject of anything like scientific study, let us turn to his *magnum opus, The Principles of Sociology,* and see how, and with what results, he puts his theory of study into practice. This immense work, full of encyclopædic detail as it is, contains certain general and comparatively simple conclusions, which can with sufficient clearness be expressed in a short summary, and which are typical of the character and the contents of Mr. Spencer's sociology as a whole. These general conclusions constitute in

outline the entire history of human progress from
the dawn of man's existence to the industrial civilisation of to-day.

The determining factors in all social phenomena are, says Mr. Spencer, primarily of two kinds—the *"external"* and the *"internal."* The former consist of some of the various physical circumstances in which each community or collection of men is placed ; the latter consist of the characters and constitutions of the men themselves. In the history of each community the chief of the external factors are these : the climate of the region which the community occupies ; the cultivability of this region; its geological and geographical character ; the way in which the fauna and flora natural to it are distributed ; and the character of the other communities by which the community in question is surrounded. One of the first generalisations, says Mr. Spencer, to which social science leads is this—that progress can begin only in climates and regions where the production of the necessaries of life is sufficiently easy to leave men leisure and energy available for other work ; and all progress did as a fact begin in those parts of the earth where the maintenance of life was easy.

He goes on to show, however, that the initiation of progress does not require only that the men concerned in it should inhabit a region in which the production of necessaries is easy and leaves them abundant leisure. . It is equally essential that the men themselves should possess an energetic tem-

perament, which will not suffer them to devote their
leisure to idleness, but will make it the starting-
point for some further activity. Now this energetic
temperament is the special gift of climate. So, to
a great extent, is the ease with which necessaries
are obtained from the soil; but whilst the fertility
of the soil is dependent on the climate being hot,

All the con-
quering races
came from
fertile and
bracing
regions.
the requisite energetic temperament is dependent
on the climate being dry. "*The evidence,*" says
Mr. Spencer, "*justifies this inference. . . . On
glancing over a general rain-map of the world, there
will be seen an almost continuous area, marked
'rainless district,' extending across North Africa,
Arabia, Persia, and all through Thibet and Mon-
golia ; and from within, or from the borders of this
district, have come all the conquering races of the
Old World.*"

There were
other regions
yet more
fertile, but
these were
enervating ;
and here the
inhabitants of
the former
enslaved the
weaker inhabi-
.tants of the
latter.
But the full operation of climate on human progress is not intelligible till a further climatic fact is
considered. Though in hot and dry climates the
production of necessaries is easy, in climates that
are hot and moist their production is still easier.
It is these last that are really the gardens of the
world, and that offered to primeval man the easiest
and most attractive homes. The original inhabitants,
however, of these favoured localities not only
profited by their conditions, but also ultimately
suffered from them. Whilst the fertility of their
habitat pampered them, its moisture destroyed their
energy ; and in process of time they were subjugated
by other races, who, cradled ' in drier climates,

retained their energy unimpaired. In this natural descent of the stronger races on *"the richer and more varied habitats"* of the weaker, and the consequent super-position of one race over another, we see the origin of slavery, and of all the ancient civilisations that reposed upon it.

We have here the three essential elements to the union of which primarily all human progress has been due : firstly, a race remarkable for its active energy ; secondly, the appropriation by this race of some richer habitat than its own ; and thirdly, the possession by it of an inferior race, as subjects, who are ready to work for its benefit, and are capable, when coerced and directed by it, of producing wealth indefinitely greater and more varied than they would or could have produced had they been left to their own devices.

And here we are brought to the threshold of a new order of facts. Industrial production, which is the basis of all civilisation, is not, says Mr. Spencer, started on its progressive career by the sudden orders of any one remarkable man, but by the spontaneous action of certain natural causes. It must first be observed that its general character and its progress are always found to depend on the same thing. They depend on the division of labour. This, as Mr. Spencer says, developed in varying degrees, is the salient characteristic of every civilisation in the world. To what, then, is the division of labour, in the first instance, itself due? This is the opening question asked by Adam

Again, division of labour, on which industrial progress depends, was caused by the differences in the products of different localities,

Smith in his *Wealth of Nations;* and he seems to regard it as one which is more or less mysterious and recondite. The answer which he himself suggests is, that there exists in man "*a natural propensity to truck, barter, and exchange.*" The answer given by Mr. Herbert Spencer is a curious illustration of how far, since the days of Adam Smith, social science has progressed.

Mr. Spencer shows us that the origin of the division of labour was no special propensity mysteriously innate in man. Its origin was the natural diversity of the various districts inhabited by the groups of men who originally took part in it. Thus "*some of the Fiji Islands,*" he writes, "*are famous for wooden implements, others for mats and baskets, others for pots and pigments—unlikenesses between the natural products of the islands being the causes. . . . So also . . . the shoes of the ancient Peruvians were made in the provinces where aloes are most abundant, for they were made of the leaves of an aloe called 'maguey.' The arms were supplied* which led to *by the provinces where the materials for making* the localisation of industries. *them were most abundant.*" Division of labour, in short, was primarily a localisation of industries, caused by the fact that a number of man's different needs were each supplied most easily by industry in some different locality.

By means of this explanation of the origin of the division of labour, Mr. Spencer proceeds to explain, in a way which would have astonished Adam Smith still more, other social phenomena of a kind which

seem wholly different. He proceeds to show us that Book 1
Chapter 2 though increased production of commodities was the
chief direct result of the localisation of industries, The locali-
sation of indus- certain by-products resulted from it also, whose tries in its turn effects were not less important. These by-products led to road-
making : were roads. In the localisation of industries, he
says, we have the true origin of road-making. The
fact of industries being widely separated in place,
required a constant interchange of the various sorts
of goods ; and the carriage of these goods to and fro
between the same points first produced tracks, such
as those made by animals, then paths, and at last
regular roads. But to facilitate the movement and
interchange of goods is not the only, or the highest,
though it may be the first, function of roads. Roads
facilitate two things of a yet more interesting and roads
made possible character—the movement of ideas and the central- the central- isation of authority. They form, in fact, the great isation of
authority and physical basis of civilised human government, and interchange of
ideas. of the development of the human intellect.

These examples of Mr. Spencer's conclusions Next, as to
men's natural will be sufficient to show how he studies the pheno- character, mena of social progress in so far as they are the which is the
other cause of result of what he calls "*the external factors*"— progress, climate, locality, and the character of the other races
with which each race that is studied happens to have
been brought in contact. Let us now turn to what he
calls the "*internal factors*," and consider the phenomena of progress which he explains by reference to
these. He helps us here by providing us with a
summary of his own, in which he calls the attention

of his readers to the most important of his own conclusions arrived at in preceding chapters as to this
section of his subject.   Having reminded us of how
he started with the "*external factors*," and how he
had shown the ways—namely those we have just
glanced at—in which they co-operated to produce
civilisation, "*our attention*," he proceeds, "*was then
directed to the internal factors*"; and what he had
to tell us, he says, about the internal factors was as

follows : "*An account was first given of ' Primitive
Man—physical,' showing that by stature, structure,
strength . . . he was ill fitted for overcoming the
difficulties in the way of advance.   Then examination of ' The Primitive Man—emotional' led us to
see that his imprudence and his explosiveness, re-
strained but little by sociality and the altruistic
sentiments, rendered him unfit for co-operation.
And then, in the chapter on ' Primitive Man—intellectual,' we saw that while adapted by its active and
acute perceptions to the needs of a wild life, his type
of mind was deficient in the faculties required for
progress in knowledge.*"   Then, having referred to
the long explanation given by him of the rise of
man's religious belief, Mr. Spencer goes on to say
that these primitive human characteristics constitute
the internal factors, with which sociology starts, and

till it was
gradually im-
proved by the
evolution of
marriage and
the family—
especially of
monogamy.

that the business of this science is to explain the
evolution of all those subsequent "*phenomena
resulting from their combined actions.*"   Of these
phenomena the chief, he says, are the following—
monogamy as evolved from polygamy, polyandry,

and promiscuity; the higher family affections as developed by the monogamous family ; and governmental and social organisation as developed in two ways—by the conduct essential to war and the conduct essential to industry. His conclusions, so far as possible, shall be given in his own words.

To begin with marriage : in the earlier stages of society nothing resembling it existed. The nearest approach to a family was the mother and such children as could be kept alive without the help of the father ; and as the children grew up, this rudimentary group dissolved. But "*from families thus small and incoherent*" there naturally and inevitably arose, in accordance with the tendency to variation by which the human units are characterised, and which is the basis of all evolutionary selection, "*families of divergent types*"—families founded on unions of which some were more lasting than others, of which some were unions between one mother and many fathers, some between one father and many mothers, and some between one father and one mother. This last-named type of union, and the family life resulting from it, had many practical advantages, such as the production of closer bonds between the several members of the family, and consequently the practice between them of more efficient co-operation. Accordingly, no sooner did monogamous groups appear than they exhibited a tendency to survive in the social struggle for existence ; and monogamy affords, with the affections that have grown up under its shelter, the type

Marginal notes: Monogamy represents the survival of the fittest kind of social union.

It developed the affections and the practice of efficient co-operation.

of marriage and family that prevails amongst the most advanced races of to-day.

Next, as to the phenomena of governmental and social organisations : these arise only with the formation of groups larger than the family—of groups which we call communities, or nations, or social aggregates ; and we have to consider how these larger groups rose out of the aggregation of the smaller. The process is explained, says Mr. Spencer, by the same few *"internal factors."* The nation sprang from the family by the following inevitable stages. Let us take any family group, sufficiently coherent to live together as a single household, and supporting itself on the produce of the land that surrounds its dwelling. Whilst this group is small, the acreage will be small also, which, as ploughland, hunting-ground, or pasture, is required to supply its wants ; and each member of the group can easily reach his work, starting from the common home, and coming back to it in the evening. But as children grow, and children and great-grandchildren

multiply, the land required by the household correspondingly grows in extent, and at last becomes so large that the whole of it cannot be utilised by a body of men living on the same spot. Hence, as Mr. Spencer expresses it, *"a fission of the group is necessitated"* ; and this process is repeated till there are a multitude of groups instead of one. These groups, says Mr. Spencer, constitute the raw material of the nation. The nation is formed *" by the recompounding of these units once again."*

And how is this process of "*recompounding*" accomplished? Mr. Spencer answers it is accomplished by one means only, and that is the co-operation forced on them by war for some common interest. Other tribes threaten to attack their territory, or they are desirous of appropriating the territory of other tribes. Separately they are powerless. The only course open to them is to band themselves together and submit themselves to a common leader. In cases where such wars are short, as observation of savage tribes shows us, the rudimentary nation with its rudimentary discipline dissolves and disappears as soon as the wars are over; but when the state of warfare is prolonged by the rivalry of other societies, the military leadership develops into a permanent centralised authority; and from this military government, with its "*coercive institutions,*" national existence and all forms of government spring.

Book I
Chapter 2

and the recompounding of these groups, for purposes of defence or aggression, formed the nation:

all government being in its origin military.

And here Mr. Spencer's argument takes a new departure and carries us on to the point where we shall be compelled to leave it. As governments and civilisations have advanced, he says, they have taken two forms—that in which the original military element still continues to preponderate, and that in which the military element becomes gradually subordinate to the industrial. "*The former,*" he says, "*in its developed form is organised on the principle of compulsory organisation, whilst the latter in its developed form is organised on the principle of voluntary co-operation*"; and the latter

But as the arts of life progress, industry gradually emancipates itself from governmental control, and becomes its own master, and also forms the basis of political democracy.

amongst civilised nations always tends to supersede the former, in precise proportion as war tends to become less common. The industrial form, it may be observed, corresponds in a general way to the kinds of government commonly called "democratic"; but its emergence, says Mr. Spencer, has its most important effects in the sphere not of politics, but of economic production. Originally the conditions of industry were regulated by the dictates of the military and aristocratic ruler, as they are to-day in some savage communities, and as they partially were in France till towards the close of the last century. Under such a *régime* the very *"right to labour"* itself is regarded as belonging to the King; and he sells it to his subjects on such terms as he may choose. But as the military element in the government declines, not only does the character of governmental legislation change, but industry frees itself from governmental influence altogether. No king any longer arranges markets, fixes wages or prices, and settles what kind and quantities of commodities shall be produced. Industry becomes, as Mr. Spencer says, *" substantially independent."* He does not mean, however, that it needs no regulation. It needs as much as ever a constant and nice adjustment of the things produced to the current requirements of the community; but this adjustment is now secured not by the interference of a political ruler, but by a system which has spontaneously developed itself amongst the trading and manufacturing classes. It is a system, says Mr. Spencer, which we may call *"internuncial, through which the various structures* (i.e. *manufacturing firms, etc.*) *receive from one another stimuli or checks, caused by rises and falls in the consumption of their respective products. . . . Markets in the chief towns show dealers the varying relations of supply and demand; and the reports of these transactions, diffused by the press, prompt each locality to increase or decrease of its special functions. . . . That is to say, there has arisen, in addition to the political regulating system, an industrial regulating system, which carries on its co-ordinating function independently — a separate plexus of connected ganglia."*

We have now looked at social evolution, as the product of both those sets of causes—the *"external factors"* and the *"internal"*—by which Mr. Spencer explains it, and have followed it, under both aspects, from the earliest beginnings of progress to the dawn and development of civilisation, such as history knows it. Our account of Mr. Spencer's theory of the ascent of man and society is necessarily very incomplete ; but the various conclusions mentioned in it may be said to be exhaustively typical of the conclusions of social science as Mr. Spencer conceives of it.

Now, if we consider all these conclusions of Mr. Spencer's, we shall find them all to be conclusions about aggregates as wholes, not about parts of aggregates.

And now let us consider what the nature of those conclusions is. We shall find that they are, one and all of them, conclusions with regard to aggregates. All the phenomena with which they deal are phenomena not of individuals, not of different classes, but of masses of men, communities, races, nations, the units of which are regarded as being virtually so similar, that what is true of one is virtually true of all. This similarity certainly is not imputed to all mankind. Men are recognised as having been different in one epoch from what they become in another, and one race and the inhabitants of one climate as being different from other men differently born and circumstanced. The primitive millions who could hardly walk upright, and whose sexual relations resembled those of the animals, are distinguished from their erect successors who married and lived in families; and the strong and energetic races are distinguished from their weaker contemporaries. But each of these aggregates is regarded as a unit in itself. The conquering race which has grown vigorous in dry regions, and the inferior race enslaved by it, which has lost its strength in moist regions, are contrasted sharply with each other; but neither is made the subject of any internal division, nor treated as though the units composing it were not virtually similar. Mr. Spencer of course admits (for this is one of the fundamental parts of his philosophy) that these wholes, these aggregates, progress through a constant differentiation of their parts, different functions being performed by an increasing number of groups; but the units who compose these groups, and whom he calls the "*internal factors*," are regarded by him as being congenitally each a counterpart of the others; and their different functions and their different acquired aptitudes are

regarded as the result of different external circumstances which press into different moulds one and
the same material.   Thus when the single group
from which the nation originally springs undergoes,
as it becomes more numerous, what Mr. Spencer
calls the process of "fission," and spreads itself in
search of food over an ever-extending area, new
groups separate not because they have different
appetites, but because, having the same appetites,
they must satisfy them in different places by the
exercise of the same faculties.   Division of labour,
as we have seen, he explains in the same way ; and
not its origin only, but its latest and most elaborate
developments.   Of the manufacturing businesses of and differences
to-day, for instance, with their promoters, managers, between simi-
capitalists, and multitudes of various workmen, not happen to be
only is each business treated by him as a single differently.
unit, but each of these units, or ganglia, is a unit
which differs from the rest for accidental reasons
only, as a gardener who happens to be digging .
may differ from a gardener who happens to be
raking a walk ; and he describes the whole as "*a
plexus of ganglia connected by an internuncial
system.*"
   The use of this last phrase, and the physiological
analogy suggested by it, illustrate yet more clearly
the fact here insisted on — namely, that for Mr.
Spencer the sociologist's true unit of interest is the
social aggregate, as a whole, to the exclusion of the
individual or of the class.   The latter are merely
the ganglia, or veins, or nerves, which are nothing

except as connected with the organism to which
they belong. Each social aggregate, in fact, is a
single animal ; and whatever is achieved or suffered
by any class or individual within it, is really achieved
or suffered, in the eye of the Spencerian sociologist,
not by the class or the individual, but by that
corporate animal, the community.

Now a study of these phenomena of aggregates
is, as has been said already, valuable for speculative
purposes. It has led those who have pursued it
to a variety of important conclusions which have
largely revolutionised our conception of human
history, and of the conditions that engender civil-
But, as has
been said
already, the
social prob-
lems of to-
day arise out
of a conflict
between
different parts
of the same
aggregate ;
therefore the
phenomena of
the aggregates
as a whole do
not help us. isations or else preclude their possibility. It has
shown us human life as a great unfolding drama,
but it has hardly given us any help at all in dealing
with the practical problems that belong to our own
day ; and the reason of this, which has already been
stated generally, must be apparent the moment we
consider what these practical problems are. Their
general character is sufficiently indicated by such
familiar antitheses as aristocracy and democracy, the
few and the many, rich and poor, capital and labour,
or, as Mr. Kidd puts it, collectivists and the
opponents of collectivism. In other words, the
social problems of to-day—like the social problems
of most other periods—are problems which arise
out of the differences between class and class.
That is to say, they depend on, and derive their
sole meaning from phenomena which are not refer-
able to the social aggregate as a whole, but which

are manifested severally by distinct and independent parts. The social aggregate, when regarded from this standpoint, is no longer a single animal, whose pains or pleasures reveal themselves in a single consciousness. It is a litter of animals, each of which has a consciousness of its own, and, together with its consciousness, interests of its own also, which are opposed to those of the others, instead of coinciding with them. Book I
Chapter 2

And now let us consider more closely out of what this opposition arises. Mr. Spencer, as we have seen, in our rapid survey of his arguments, lays great stress on the fact that as men rise into aggregates, they do so only on condition of submitting themselves to governors, military in the first place, and at a later stage civil. The truth, however, which he thus elaborates, whatever may be its speculative importance, fails to have any bearing on any practical problem, because it is not a truth about which there has ever been any practical disagreement. Aristocrat, democrat, and socialist all agree that there must be orderly government of some sort, and official governors to administer to it. The point at issue between them is not whether some must govern and others submit to be governed, but how the individuals who perform the work of government shall be chosen, and what, apart from their official superiority and authority, shall be their position with regard to the rest of the community. Why should they enjoy any special social advantage? Or if they are to enjoy it, why should they be usually The conflict
between the
parts of the
aggregate
arises from in-
equalities of
position,

Book I
Chapter 2

drawn from a small privileged class, and not from the masses of the community, sinking to the general level again when their tenure of office terminates? Such are the questions proposed by one party; whilst the other party replies by contending that the limited class in question can alone supply governors of the required talents and character. Of this clash of opinions and interests, which is as old as civilisation itself, though in each age it assumes some different form, Mr. Spencer's social science necessarily takes no cognisance, because the parts of each social aggregate have for him no separate existence.

of which Mr. Spencer's sociology takes no account.

The same criticism applies to his treatment of economic production. He explains, as we have seen, the origin of the division of labour, showing how "*unlikeness between the products of different districts*" inevitably led to "*the localisation of industries*," turning one set of savages—to use his own example—into potters, another into makers of baskets. But here again we have a truth which, whatever its speculative interest, has no bearing on any practical problem; for no one denies that division of labour is necessary, nor do any of the difficulties of to-day turn upon its remote origin. Socialists and individualists are alike ready to admit that different men must follow different industries. The point at issue is why, within the limits of the same industry, different men pursue it on different levels, some being masters and capitalists, some being labourers and subordinates. Here, just as in the sphere of political and military government, we have one class defending its existing position and privileges, and another class attacking or questioning them ; and it is out of circumstances such as these, thus briefly indicated, that the practical social problems of the present day arise.

Now the question at the bottom of these can be reduced to very simple terms. If all members of the community were content with existing social arrangements, it is needless to say there would be no social problems at all. Such problems are due entirely to the existence of persons who are not contented, and who desire that certain of these arrangements should be changed. It will be seen, accordingly, that the great and fundamental question which, as a practical guide, the sociologist is asked to answer, is whether or how far the changes desired by the discontented are practicable ; and the first step towards ascertaining how far the arrangements in question can be turned into something which they are not, is to ascertain precisely how they have come to be what they are.

But this way of putting the case is still not sufficiently definite. Mr. Spencer himself has put it in somewhat similar language ; and yet in doing so he has missed the heart of the problem. Mr. Spencer's speculative gaze, travelling over the past and present, sees one generation melting like a cloud into another, and takes no note of the individuals that compose each. The practical sociologist must adopt a very different method of observation. He must remember that practical problems arise

*Book I
Chapter 2*

Social problems arise out of the desire of those whose positions are inferior to have their positions changed ;

and the practical question is, is the change they desire possible?

and become practical, not in virtue of their relation
to mankind generally, but in virtue of their relation
to each particular generation that is confronted by
them ; and a particular generation in any given
community, and the different classes into which the
community is divided, are made up respectively of
particular men and women.   In asking, therefore,
how the social arrangements we have been considering have come to be what they are, we must not ask
in vague and general terms why a portion of the
social aggregate occupies a position which contents
it, and another portion a position which exasper-
ates it ; but we must consider the individuals of
which each portion, at any given time, is composed,
and begin the inquiry at the point at which they
begin it themselves.   " Why am I—Tom or Dick
or Harry—included in that portion of the aggregate
which occupies an inferior position ?   And why are
these men—William or James or George—more
fortunate than I, and included in the portion of
the aggregate which occupies a superior position ?"
To this question there are but three possible
answers.   The inferior position of Tom or Dick
or Harry is due to his differing from William or
James or George in external circumstances, which
theoretically, at all events, might all be equalised
—such, for example, as his education ; or it is due
to his differing from them in certain congenital
faculties, with respect to which men can never
be made equal—as, for example, in his brain power
or his physical energy ; or it is due to his differing

To answer this
question we
must examine
into the causes
why such and
such indi-
viduals are in
inferior, and
others in
superior posi-
tions.

from them in external circumstances which have
arisen naturally from differences in the congenital faculties of others, and which, if they could be equalised at all, could never be equalised with anything like completeness—such, for example, as the possession by William and James and George of leisured and intellectual homes secured for them by gifted fathers, and the want of such homes and fathers on the part of Tom and Dick and Harry.

The first question, accordingly, which we have to ask is as follows. Taking Tom or Dick or Harry as a type of those classes who happen to occupy an inferior position in the aggregate, and comparing him with others who happen to occupy superior positions, we have to ask how far he is condemned to the inferior position which he resents by such external circumstances as conceivably could be equalised by legislation, and how far by some congenital inferiority of his own, or circumstances naturally arising out of the congenital inferiority of others. Or we may put the question conversely, and ask how William and George and James have come to occupy the positions which Tom, Dick, and Harry envy. Do they owe their positions solely to unjust and arbitrary legislation, which a genuinely democratic parliament could and would undo? Or to exceptional abilities of their own, of which no parliament could deprive them? Or to advantages secured for them by the exceptional abilities of their fathers, which no parliament could interfere with, or, at all events, could abolish, without

Are inequalities in position due to alterable and accidental circumstances?

Or are they due to congenital inequalities which no one can ever do away with?

entering on a conflict with the instincts of human nature, and interfering with the springs of all human action?

Social inequali-
ties are partly
due to circum-
stances ;
Now that external circumstances of a kind, easily alterable by legislation, have been, and often are, responsible for many social inequalities, is a fact which we may here assume without particularly discussing it. The inquiry, therefore, narrows itself still further, and resolves itself into this: Do the congenital superiorities or inferiorities of the persons, or of parents of the persons, who at any given time are occupying in the social aggregate superior and inferior positions, play any part in the production of these social inequalities at all?

This question must plainly be the practical sociologist's starting-point; for if social inequalities are due wholly to alterable and artificial circumstances, social conditions are capable, theoretically, at all events, of being equalised; but if, on the other hand, inferior and superior positions are partly, at all events, the result of the congenital inequalities of individuals, over which no legislation can exercise the least control, then a natural limit is set to the possibilities of the levelling process; and it is the business of the sociologist, if he aspires to be a

but most
people will
admit that con-
genital in-
equalities in
talent have
much to do
with them.
practical guide, to begin with ascertaining what these limits are. Are, then, the congenital inequalities of men a factor in the production of social inequalities, or are they not?

Now to many people it will seem that even to ask this question is superfluous. They will regard

it as a matter patent to common sense that men's Book 1 Chapter 2
congenital inequalities are to a large extent the
cause, in every society, of such social inequalities Why then
as exist in it; and they will possibly say that it is insist on this fact?
a mere waste of time to discuss a truth which is so Because this fact is pre-
self-evident. It happens, however, that the more cisely what our
obvious it seems to be to common sense, the more contemporary sociologists
necessary it is for us to begin our present inquiry ignore.
with insisting on it; and the reason is that, in spite
of its being so obvious, the whole school of contemporary sociologists, with Mr. Spencer as their head,
base their whole method of sociological study on a
denial of it. By their method of dealing with social
aggregates only, they deny not only the influence,
but even the existence of congenital inequalities,
and endeavour to explain them away as an illusion
of the unscientific mind. They admit, indeed, as
our quotation from Mr. Spencer showed, that the
primitive man was congenitally different from man in
later ages. They admit that the individuals reared
in a dry climate, who formed the conquering aggregates, were congenitally different from the individuals
reared in a moist climate, who formed the enslaved
aggregates; but they absolutely refuse to take any
account whatever of the congenital inequalities by
which individuals within the same aggregate are
differentiated.

In order to show the reader that such is literally
the case, we need not rely merely on such inferences
as have just been drawn from the manner in which
Mr. Spencer applies his method, and from the

4

Book 1
Chapter 2

as Mr. Spencer
shows us by
his distinct
admissions and
assertions, as
well as by the
character of
his conclusions. general character of his conclusions.   We have the direct evidence of his own categorical statements. Let us turn again to the criticism with which, as we have already seen, he prefaces his whole series of sociological writings, and which may be taken as his fundamental profession of faith—his criticism, namely, of what he calls "*the great-man theory*," his rejection of it as being a theory which would render all social science impossible, and his enunciation of the theory which he contends must take its place.   It may seem to some readers that his rejection of the great man as a *vera causa* which will explain social phenomena amounts to no more than a rejection of that exaggerated view of history which expresses itself in the works of writers such as Froude and Carlyle, and which vaguely attributes all the progressive changes of humanity to the personality of rulers, of political and military autocrats —such as Henry VIII., Cromwell, and Frederick the Great of Prussia.   And indeed, to judge by Mr. Spencer's language, it is this exaggerated view which has been most frequently present in his mind, as we may see by referring to the passage already quoted, which concludes his demonstration that the "*great-man theory*" is false.   With the sole exception, he says, of the military struggles of primitive tribes, "*new activities, new institutions, new ideas, unobtrusively make their appearance, without the aid of any king or legislator; and if you wish to understand the phenomena of social evolution, you will not do it should you read yourself*

blind over the biographies of all the great rulers
on record, down to Frederick the Greedy and
Napoleon the Treacherous."

But Mr. Spencer, in rejecting the great "*ruler* His condemnation of the
great-man
theory is a
removal of all
congenital in-
equalities from
his field of
study ;
and legislator" as a factor in social evolution unworthy of the attention of the sociologist, is really
rejecting a great deal else besides. He is really
rejecting every inequality in capacity by which a
certain number of men are differentiated from, and
raised above others. In order to show that such is
the case, we will avail ourselves of his own words.
We will, then, start with one casual remark out of
many, in which Mr. Spencer, forgetting his own
theories, slips into a method of observation truer than
the one he advocates. "*Men,*" he writes in his *Study*
of Sociology, "*who have aptitudes for accumulating*
observations are rarely men given to generalising;
whilst men given to generalising are commonly
men who, mostly using the observation of others,
observe for themselves less from love of particular
facts than from the desire to put such facts to use."
Nothing can be clearer than the distinction here
drawn. It is one of great importance in the
elucidation of many social problems ; and it deals not
with the likeness, but with a congenital difference,
which exists between men belonging to the same
social aggregate. But now let us compare this
with another passage, in which Mr. Spencer, re-
turning again to his theory, explains how members
of the same aggregate are to be treated by any
sociologist who would claim to be a man of science.

"*Amongst societies of all orders and sizes,*" he writes, "*sociology has to ascertain what traits there are in common, determined by the common traits of human beings; what less general traits, distinguishing certain groups of societies, result from traits distinguishing certain races of men; and what peculiarities in each society are traceable to the peculiarities of its members.*" This is clumsily expressed; but its meaning, which is quite obvious, may be seen by taking, as a typical society, that of England. The sociologist, in explaining English society, will have to consider, according to Mr. Spencer, first, what traits Englishmen have in virtue of being human creatures; secondly, he will have to consider what traits they have in virtue of being Europeans, not Orientals; and, thirdly, he will have to consider what traits they have in virtue of being Englishmen, not Frenchmen or Germans.

The reader will at once perceive the contrast between the spirit of these two passages. In the former Mr. Spencer notes, with great penetration and accuracy, a most important point of difference between two sets of men belonging to the same society. In the latter he deals with societies as single bodies, the members of which possess no personal traits whatever, except such as they all possess alike; and all the traits in which they differ from one another, such as the one just alluded to, of necessity disappear from the field of vision altogether. Should any doubt as to the matter still remain in the reader's mind, it will be dispelled by

and he actually defines an aggregate as being composed of *approximately equal units.*

the quotation of one further passage. "*A true* *social aggregate*," he says ["*as distinct from a mere large family*], *is a union of like individuals, in-dependent of one another in parentage, and approxi-mately equal in capacities.*"

Here is the case stated with the most absolute clearness. All congenital inequalities, as was said just now, between the various individuals who make up the aggregate are ignored; and it is upon this hypothesis of approximately equal units, acted on by different external circumstances, that he attempts to build up his whole system of sociology. He is, indeed, little as he himself may suspect it, reproducing in another form the error of Karl Marx and the earlier of the so-called "scientific socialists," who maintained that all wealth was the product of common or average labour, measured by time, and that hour for hour any one labourer necessarily produced as much wealth as another. The socialists of to-day are already beginning to see that this monstrous, though ingeniously advocated, doctrine is untenable as the foundation of economics ; and yet, strange to say, a doctrine strictly equivalent to it forms the accepted foundation of con-temporary social science. That science starts with the hypothesis of approximately equal units, and ignores the congenital differences between the individuals who compose the aggregate. We shall find it to be ultimately from differences of this kind that all the practical problems which beset civilisa-tion spring, and that the inability of the modern

His failure, and that of others, as practical sociologists, arises from their building on this false hypothesis.

sociologists, complained of by Mr. Kidd and Professor Marshall, to throw on these problems any definite light is simply the natural and inevitable result of excluding the differences in question altogether from their scientific purview.

We will, in the next chapter, consider the whole range of arguments used by Mr. Spencer and others in justification of this error.

CHAPTER III

GREAT MEN, AS THE TRUE CAUSE OF PROGRESS

I T is evident that an error of the kind now in question does not represent the carelessness of the untrained thinker. It is nothing if not deliberate ; and indeed Mr. Spencer admits that it is altogether in opposition to the opinions which men naturally hold. Accordingly, the arguments by which he and his followers justify it, and have actually imposed it on all the sociological thinkers of their generation, require, before we reject them, to be examined with the utmost care.

The ignoring of natural inequalities is a deliberate procedure. Let us see how it is defended.

Let us examine Mr. Spencer's defence of it.

Let us then turn our attention once again to the grounds on which Mr. Spencer refuses to admit the great or exceptional man as a true factor in the production of social change. If the reader will reflect upon the account that has been already given of Mr. Spencer's arguments in connection with this point, he will find that Mr. Spencer rejects the great man for two reasons, which are not only distinct, but are, when interpreted closely, not entirely consistent with each other. One of these reasons is that the great, or exceptional man does

He defends it in two ways ;

(1) by saying that the great man does not really do what he seems to do ;

ARISTOCRACY AND EVOLUTION

"*if there is to be anything like a real explanation*"

"*proximate initiator*"—changes, to quote an example

"*in the aggregate of social conditions out of which

Spencer's second contention is expressed in the

of ruler and ruled which was good at one time is

begin to give origin to new activities, new ideas, all of which unobtrusively make their appearance without the aid of any king or legislator."

It will be necessary to deal with these two contentions separately; and we will begin with the second, as set forth in the words just quoted. We shall find it valuable as an example of that singular confusion of thought by which all the reasoning of our sociologists with regard to this question is vitiated. Mr. Spencer speaks of an *"immense error"* which he is pointing out and correcting. The *"immense error"* in reality is to be found in his own conception. It is hard to imagine anything more arbitrary and more gratuitously false than the contrast which he here draws between the actions of men in primitive war, for the success of which he admits a great leader to have been essential, and their various actions and activities as manifested in peaceful progress, which, he contends, neither require leadership nor exhibit traces of its influence. We are at this moment altogether waiving the question of how far the great leader, when he is the proximate cause of the military successes of his tribe, is their cause in any deeper sense. It is enough for us now to take Mr. Spencer's admission that the leader is really the cause, in some sense or other, of the social changes connected with early warfare; and, keeping to this sense, let us consider in what possible way less causality can be attributed to the actions of great men and leaders in the sphere of peaceful progress.

He admits that the great man does do something exceptional in war;

but denies that he does anything exceptional in the sphere of peaceful progress.

Book 1
Chapter 3

But how does
the great man
fulfil his
function in
war? By
ordering
others.

"*A primitive society,*" if it is to become powerful in war—this Mr. Spencer admits—must have a great leader to direct it. But what precisely is it that such a leader is and does ? Such a leader leads, because he is one mind or personality impressing for the moment its superior qualities on many minds or personalities. He supplies the fighting men of his society with an intelligence not their own—often with a courage, a presence of mind, and a resolution. He dictates to them the directions in which their feet are to carry them ; the manner in which they are to group themselves ; the movements of their hands and arms. He gives the word, and a thousand men dig trenches. He gives the word again, and a thousand men wield swords ; now he makes them advance ; now he makes them halt ; and the measure of his greatness as a leader is to be found in those results which, by directing the action of all these men, he elicits from it.

And now from the triumphs of war let us turn to those of peace. "*These,*" says Mr. Spencer, "*unlike the former, make their appearance unobtrusively, without the aid of any king or legislator.*" It may, no doubt, be true that they do appear unobtrusively in the sense that they are not accompanied by trumpets and drums and tom-toms. A factory for the production of toffee, or of trimmings for ladies' petticoats does not require an Ivan the Terrible to direct it, nor are Mr. Spencer's sentences as he writes them punctuated by discharges of artillery. But if the essence of kingship and leader- ;

ship is to command the actions of others, the larger part of the progressive activities of peace, and the arts and products of civilisation, result from and imply the influence of kings and leaders, in essentially the same sense as do the successes of primitive war, the only difference being that the kings are here more numerous, and though they do not wear any arms or uniforms, are incomparably more autocratic than the kings and czars who do.

As a particularly clear illustration of this important truth, let us take Mr. Spencer himself, and place him before his own eyes as an autocratic king or ruler. In certain respects he is so; and it is only because he is so that he has been able to give, through his books, his thoughts and theories to the world. For let us examine any one of his volumes and consider what it is, in so far as it differs from any other volume—let us say from a treatise on the cutting of trousers, or an attack on the Spencerian philosophy—which is printed in similar type on pages of the same size. It differs solely in the order in which the letters have been arranged by the hands of the compositors ; and its value as a work of philosophy consequently depends altogether on a certain complicated series of movements which the hands of the compositors have made. And how has this prolonged series of minute movements been secured ? It has been secured by the fact that Mr. Herbert Spencer, through his manuscript, has given the compositors a prolonged series of orders, which their hands, day after day, have been obliged to obey

Book I
Chapter 3

The great man, in peace, does precisely the same thing.

Mr. Spencer, for example, orders the compositors who put his books into type.

passively.   He has been as absolute a master of
all their professional actions as ever was the most
arbitrary general of the professional actions of his
soldiery; and there is absolutely no difference in
point of command and obedience between the
compositors who, at Mr. Spencer's bidding, put into
type the words " *homogeneity* " and " *the Unknow-
able*," and the Guards who charged the French at
the bidding of the Duke of Wellington.

The inventor
orders the men
by whom his
inventions are
manufactured.
Precisely the same thing is true of all scientific
inventions—not indeed of inventions as mere ideas
and discoveries, but of inventions and discoveries
applied practically to the service of civilisation.
The mere discovery of certain properties belonging to material substances, or the thinking out
of some new machine or process, may be the
work of one man, who has command over nobody
except himself.   But the moment he proceeds
to make his machine or process useful—to apply it
to the purpose of actual business or manufacture—he
is obliged to secure for himself an entire army of
mercenaries, who act under his orders in precisely
the same way as soldiers act under the orders of the
military leader, or as the compositors act under the
orders of Mr. Spencer.   When the electric telegraph
was supplemented by the invention of the telephone,
telephones were produced, and could have been
produced, only by a multitude of men performing a
series of manual actions which were different in detail
from anything they had performed before, and which,
if it had not been for the inventor, would never

have been performed at all. They filed or they
cast pieces of metal into new shapes ; with these pieces of metal they connected in new order pieces of other materials, such as wood and vulcanite, the shape of these last being new and special also ; and every piece of material shaped or connected with another piece was the exact resultant of so many manual movements made in passive obedience to the inventor's autocratic orders. It was only because his orders were obeyed with such humble fidelity and completeness that these movements resulted in telephones, enriching the world with a new con- venience, and not in the old-fashioned telegraphic machines, or in penholders, or vulcanite inkstands, or even in useless heaps of shavings and brass filings. And the same is the case with every inven- tion or contrivance which has helped to build up the fabric of modern material civilisation.

Civilisation, however, even in its most material sense, does not consist of contrivances and inventions only. "*The one operation,*" says Mill, "*of putting things into fit places . . . is all that man does, or can do, with matter. He has no other means of acting on it than by moving it.*" But valuable as this formula is, it is not sufficiently comprehensive ; for there is another economic process which, to the ordinary mind at all events, is hardly suggested by such a phrase as "*to move matter.*"

The process referred to consists in the moving of men. What is meant by the distinction here drawn is this—that the industrial efficiency of a community does not depend solely on the muscles of the manual workers being given a right direction, so that they shall shape material objects in such and such a way ; but it depends also on the movements which are prescribed to the men, being prescribed to the men best fitted to perform them, and being prescribed to them in such order that when each movement has to be made, the men told off to make it shall be ready to make it at the moment. Here we see part of the secret of the success of the great contractor.

The importance of these considerations becomes all the clearer to us when we reflect on the fact that the mere production of commodities, and the production of the means of production, form but a part of the processes which advance, maintain, and indeed constitute civilisation. A part almost equally large consists in the rendering of various personal services, which often, no doubt, involve the utilisation of improved appliances, but which almost as often are neither more nor less than the performance of actions of a simple and ordinary kind, the merit and demerit, the wastefulness or the economy of which depend on their being performed with absolute punctuality and despatch. A good example of this is the case of a large hotel. Whether a large hotel is carried on at a profit or at a loss depends almost entirely on this question of personal management. The success of a successful manager does not depend on his capacity for inventing new methods of waiting, of cooking, or of making beds. It depends on his capacity for organising his staff of cooks, waiters, and chamber-maids. This is well expressed by that most significant American saying, "He's a smart man, but he couldn't keep a hotel"; the meaning being that one of the most important, and at the same time one of the rarest faculties required for maintaining a complicated civilisation like our own is the faculty by which, given a number of tasks, one man governs a number of men in the act of co-operatively performing them.

Book I
Chapter 3

Examples of this kind might be indefinitely multiplied, but those just adduced are quite sufficient to prove the sole point insisted on at the present moment—namely, that whatever be the part (and Mr. Spencer admits it to be "*all-important*") which the great man plays as a leader in primitive warfare, a part precisely similar in kind is played by other great men in the peaceful processes, and, above all, in the progress of civilisation.

All these men resemble the great military commander, and if the latter is a social cause, so are the former.

And now, having dealt with this point, let us turn to Mr. Spencer's other contention—his contention namely that, whatever the part may be, and however seemingly important, which the great man plays in producing social changes, he is, in any case, nothing but their "*proximate initiator*";—that "*they have their chief cause in the generations he descended from*";—and that if there is to be anything like a real and scientific explanation of them, it must be sought in the aggregate of conditions out of which both he and they have arisen, and not in the great man's personality as revealed to us by any

Next, as to the contention that the great man is the proximate cause only, and not the true cause—

records of his life, or by any analysis of his peculiar faculties.

We have already seen in a general way how this feat of merging the great man in "*the aggregate of*
this, as Mr.
Spencer, and
three popular
writers of to-
day show us,
conditions out of which he has arisen" is performed by Mr. Spencer himself. Let us now turn for a moment to three other writers who, though differing from him as to certain of his conclusions, have with regard to this particular point done little else than popularise and apply his teaching.

"*It needs only a little reflection,*" writes Mr. Kidd, "*to enable us to perceive that the marvellous accomplishments of modern civilisation are primarily the measure of the social stability and social efficiency, and not of the intellectual pre-eminence of the peoples who have produced them. . . . For it must be remembered that even the ablest men amongst us, whose names go down to history connected with great discoveries and inventions, have each in reality advanced the sum of knowledge by only a small addition. In the fulness of time, and when the ground has been slowly and laboriously prepared for it, the great idea fructifies and the discovery is made. It is, in fact, the work not of one, but of a great number of persons. How true it is that all the great ideas have been the products of the time rather than of individuals may be the more readily realised when it is remembered that, as regards a large number of them, there have been rival claims put forward for the honour of authorship by persons who, working quite independently, have arrived at like results almost simultane-*

ously. Thus rival and independent claims have been
made for the discovery of the differential calculus . . .
*the invention of the steam engine, . . . the methods
of spectrum analysis, the telegraph, the telephone,* ·
as well as many other discoveries." And then
Mr. Kidd proceeds to quote with approval the
following sentence from an essay which was written
by an American socialist, Mr. Bellamy; and the
sentence has been repeated with solemn and
triumphant unction in half the socialistic books
which have been given to the world since.
*"Nine hundred and ninety-nine parts out of the
thousand of every man's produce are the result of
his social inheritance and environment."* *"This
is so,"* remarks Mr. Kidd, *"and it is, if possible,
even more true of the work of our brain than of the
work of our hands."* To these passages we must
add one from Mr. Sidney Webb, who is, intellectually, a favourable example of a modern English
socialist. Referring to the socialistic proposal that
all kinds of workers, no matter what their work,
should be paid an equal wage, *"this equality,"* he
says, *"has an abstract justification, as the special
ability or energy with which some persons are born
is an unearned increment due to the effect of the
struggle for existence upon their ancestors, and
consequently, having been produced by society, is as
much due to society as the unearned increment of
rent."* -

Here we have then, in the words of these four
writers, Mr. Spencer, Mr. Kidd, Mr. Bellamy, and

5

Mr. Sidney Webb, the case against the great man set fully before us ; and we may accordingly proceed to analyse it. We shall find that it divides itself into four separate arguments, which are constantly recurring in some form or other in all the works of our modern sociological writers, and especially in the works of those who are democratic or socialistic in (1) That every first discovery involves all that have gone before it ; their sympathies. Firstly, there is the argument that in any advanced civilisation not one of the improvements made during any given epoch would have been possible if a variety of other improvements and the accumulation of various knowledge had not gone before it ; and that thus the man who is called the inventor or author of the improvement is merely the vehicle or delegate of forces outside (2) that the discoverer's ability itself is the product of past circumstances ; himself. Secondly, there is the argument that the inventor or author of the improvement, even if we attribute to him some special ability of his own, is in respect of his own congenital energies merely the product and expression of preceding generations and circumstances. Of the four arguments in question, these are the most important ; but they (3) that often the same discovery is made by several men at once ; are constantly reinforced by two others. One is drawn from the fact that several independent workers often arrive simultaneously at the same (4) that the difference between the great and the ordinary man is slight. discovery. The other is drawn from the fact— or what is alleged to be the fact—that the interval which divides even the greatest man from his fellows, alike in respect of what he is and of what he accomplishes, is really extremely slight, and not worth considering.

For convenience' sake, we will deal with these two latter arguments first, and put them out of the way before we approach the others. We will begin with the argument drawn from the fact that the same discovery is often made simultaneously by independent workers. This would perhaps hardly be worth discussing if it were not used so constantly by such a variety of serious writers. The fact is true enough, but what is the utmost that it proves? If two or three men make the same discovery at once, this does not prove, as it is supposed to do, that all men are approximately equal, but that two or three men, instead of one man, are greater than the rest of their fellow-workers. If three horses at a race out-distance all competitors, and pass the winning-post within the same three seconds, this does not prove that a cart-horse is as swift as the Derby favourite. As a matter of fact, that more men than one should reach at the same time the same discovery independently is precisely what we should be led to expect, when we consider what discovery is. The facts of nature which form the subject-matter of the discoverer are in themselves as independent of the men who discover them as an Alpine peak is of the men who attempt to scale it. They are indeed precisely analogous to a peak which all discoverers are attempting to scale at once; and the fact that three men make the same discovery simultaneously does no more to show that any of their neighbours could have made it, and that it is made in reality, not by them, but by

Book I
Chapter 3

Simultaneous discovery only shows that several men, instead of one, are greater than others.

their generation, than the fact that the three most intrepid cragsmen in Europe meet at last on the same virgin summit, which other adventurers had sought to scale in vain, would prove the feat to have been really accomplished by the mass of tourists at Interlaken, who had never climbed anywhere except by the Rigi railway, and whose stomachs would be turned by a precipice of twenty feet.

Let us now turn to the argument that the inequalities between men's abilities are small, that the work accomplished by even the ablest is small also, and that the exceptional man as a separate subject of study may, in the words of a writer who will be quoted presently, be in consequence " *safely neglected.*" The answer to this is that whether an inequality be great or small depends altogether on the point from which the total altitude is measured. · If a child who is three feet high, and a giant who is nine feet high, are both of them standing on the summit of Mont Blanc, the difference between the elevation of their respective heads above the sea-level will be infinitesimal ; but no one who was discussing the question of human stature would say that little children and giants were of approximately the same height. Similarly, if our object is to compare men in general with all other living creatures, no doubt the difference between the ordinary man and a microbe is incomparably greater than the difference between an ordinary man and Newton ; but if our object is to compare men with men, in relation to this or that mental capacity—let us say the capacity for scientific and mathematical discovery — the difference which separates one ordinary man from another is insignificant when compared with the difference by which Newton is separated from both of them. And it is this latter sort of difference which alone concerns the sociologist. The difference which separates men from microbes is nothing to him. And what is true of what men are, is equally true of what they do. The addition made by any one great man to knowledge may be small when compared with the knowledge, regarded in its totality, which has been gathered together by all other great men preceding him ; but it may at the same time be incalculably great when compared with the additions made by the ordinary men, his contemporaries.

It may be slight to the speculative philosopher, but to the practical man it is all-important.

Let us make this matter yet clearer by reference to one more authority, who, though endeavouring to confirm the very argument which is here being exposed, is, little as he perceives it, assassinated by his own illustrations. In Macaulay's essay on Dryden there occurs the following passage, a part of which anticipates the exact phraseology of Mr. Spencer. "*It is the age that makes the man, not the man that makes the age. . . . The inequalities of the intellect, like the 'inequalities of the surface of the globe, bear so small a proportion to the mass, that in calculating its great revolutions they may safely be neglected.*" The passage is quoted for the sake of this last simile. For those who study the human destiny as a whole—who

survey it as speculative and remote observers—the
inequalities of intellect may, it is quite true, be
neglected as safely as the inequalities of the surface
of a planet are neglected by the astronomer who is
engaged in calculating its revolutions. But because
these latter inequalities are nothing to the astronomer,
it does not follow that they are nothing to the
engineer and the geographer. To the astronomer
the Alps may be an infinitesimal and negligeable
excrescence, but they were not this to Hannibal or
the makers of the Mont Cenis tunnel. What to
the astronomer are all the dykes in Holland? But
they are all the difference to the Dutch between a
dead nation and a living one.

And the same difference, even in its most minute
details, holds good between speculative, or as we
may call it star-gazing, sociology and sociology as
a practical science; for is it not one of Mr. Spencer's
most important and interesting contentions that
these very irregularities of the earth's surface—
these lands, seas, plains, valleys, and mountains—
which, when compared with the mass of the earth,
are so absolutely inappreciable, constitute some of
the most important of the "*external factors*" of
human history and civilisation? And the same
holds good of the inequalities of the human intellect.
They may be nothing to the social star-gazer, but to
the social politician they are everything.

So much, then, for two of the most shallow
sophisms that ever imposed themselves on presumably serious reasoners. We will now turn to

those two other arguments in which the case against the great man finds its main support, and which, however misleading they may be, must be examined at greater length. In both of these the distinctly exceptional character of the great man is assumed, or at all events is not denied, but it is represented as being, if it exists, not properly the great man's own. The first argument refers it to aggregates of external conditions—the knowledge accumulated for the great man's use, the character of his fellow-citizens, who are ready to carry out his orders, and generally to what Mr. Bellamy calls his "*social inheritance and environment.*" The second argument refers it to the great man's line of ancestors, insisting that he inherits from them his own exceptional capacities, which capacities his ancestors acquired by being members of society, and of which it is accordingly contended that society is ultimately the source.

Now on both these arguments, before we consider them in detail, there is one broad criticism to be made, which applies to both equally. There is a certain sense—a remote and speculative sense— in which they are both of them quite true, and indeed are almost truisms; but for practical purposes they are either not true at all, or if true, are altogether irrelevant; and it is necessary to show the reader, by a few simple examples, that in the doctrine that statements can be at once true and not true there is no philosophical hair-splitting, and no Hegelian paradox, but merely the assertion of a fact which, when once attention has been called to it, common sense will perceive to be as obvious as it is important.

It was just now observed that the same thing can be great and not great, according to the things with which we compare it. In the same way the same statement may be true or not true, according to the nature of the discussion on which it is brought to bear. Let us take as an example those familiar statements of fact which are given in terms of averages. If the vast majority of any given population vary in height between the limits of five feet six and six feet, the statement that a man's average height is from five feet seven to five feet eight would be a truth most important to the producers of ready-made overcoats. But if half the population were two feet high, and half rather more than nine feet, to give the average stature as something like five feet seven would be for the coatmakers the most absurd misstatement imaginable, and would lead them to make, if they acted on it, garments that would fit nobody.

Let us turn from the question of the truth of a statement to the question of its mere relevance; and we can illustrate what has been said by an example equally homely. In the transference of goods by rail, these have to be sorted according to bulk, weight, shape, fragility, perishability, and so forth. In deciding which are to be sent by fast trains, and which by slow, the primary question will be that of perishability. When the perishable and the non-perishable shall have been separated, and they are being placed on the trains allotted to them, the primary questions will be those of shape, weight, and fragility. But so long as the preparatory separation is in progress, to assert that the goods possess any of these latter characteristics will be wholly irrelevant, no matter how true. Boxes of fish will not be put with book parcels because neither of them are fragile, or because they are both oblong; and each characteristic, and every classification based on it, will be either relevant or irrelevant, full of meaning or meaningless, according to what question, out of a considerable series, has to be answered at the moment by the officials who superintend the business.

And now let us go back to the two arguments that are before us; and we shall be prepared to see how, though true for the speculative philosopher, they have no meaning, or only a false meaning, for any practical man.

We will first take that which is expressed with sufficient plainness in the passage quoted from Mr. Sidney Webb, and which insists on the great man's debt to society generally, not for his external circumstances, but for his personal character and capacities. The idea involved in it is very easy to grasp. The great man's congenital superiority is an inheritance from his superior ancestors; but his ancestors would not have had it to hand on to him if they had not been forced to develop such superiorities as they possessed by exerting them in a competitive struggle

Thus the argument that the great man owes his faculties to his ancestors, and through his ancestors to the society which helped to develop his ancestors, though a speculative truism, with the great mass of their contemporaries. Thus the mass of their contemporaries formed a strop or hone on which the superior faculties of these men were sharpened ; and the great man of to-day, to whom the superior faculties have descended, owes them accordingly, not to his own ancestors only, but to the mass of inferior men who struggled with them, and were worsted in the struggle. In other words, the greatness of the exceptional man has really been produced by the whole body of society in the past ; and the results of it ought to be divided amongst the whole body of society in the present.

Now that the above line of argument has a certain kind of truth in it, it is hardly necessary to observe ; and for biologists, psychologists, and speculative philosophers generally, such truth as it possesses may no doubt be of value ; but that this truth has no relation whatever to practical life, and no applicability to any one of its problems, can be seen by considering the kind of results we shall arrive at, if, adopting the reasoning of Mr. Webb and his friends, we merely carry it out to the more immediate of its logical consequences.

leads to
nothing but
absurdities if
we apply it to
practical life.

Let us begin with their reasoning, so far as it concerns the past. If the inferior competitors who were beaten by the great man's ancestors are to be credited with having helped to produce the talents by which they were themselves defeated, and must therefore be held to have had a claim on the wealth which these talents produced, which claim has descended to the inferior majority of to-day, the same claim might be advanced by any weaker nation which, after a series of battles, succumbs finally to the stronger. In the Franco-German War the French might have said to the Germans, "You acquired by fighting with us, the faculties which have enabled you to conquer us. Your strength therefore, in reality, belongs to us, not you ; and hence justice requires that you should give us back Alsace." In the same way it might be urged that all the idle apprentices of the past have, by the warning they afforded, stimulated the industry of the industrious, and therefore in abstract justice had a claim on their earnings.

Let us now take Mr. Webb's reasoning so far as it concerns the present, and we shall find that it results in similar fantastic puerilities. If the great man of to-day owes his greatness to society as a whole, it is to society as a whole that the idle man owes his idleness, the stupid man his stupidity, the dishonest man his dishonesty ; and if the great man who produces an exceptional amount of wealth can, with justice, claim no more than the average man who produces little, the man who is so idle that he shirks producing anything may with equal justice claim as much wealth as either. His constitutional fault, and his constitutional disinclination to mend it, are both due to society, and society, not he, must suffer. And the same thing holds good of every form of economic incompetence.

The absurdity of Mr. Webb's position will be seen yet more clearly when we see how it looks when stated in the language of Mr. Bellamy. *"Nine hundred and ninety-nine parts out of the thousand of every man's produce are the result of his inheritance and his environment."* Now if this proposition has any practical application, it must mean that the whole living population—great men and ordinary men, labourers and directors of labour —who are commonly held to be the producers of the income of Great Britain to-day, really produce of it only one farthing in the pound; and hence, if we still persist in considering the proposition a practical one, we shall be forced to conclude that the whole of the living population might at any given moment stop work altogether, or fall into a trance like the Seven Sleepers of Ephesus, and the production would continue with hardly an appreciable diminution.

The same argument applies to morals; and if accepted, we should have to admit that nobody really did, or was really responsible for, anything.

Again, if the proposition has any practical bearing on economics, it must necessarily have a bearing precisely similar on morals. If a man of to-day produces only a thousandth part of what he seems to produce, it is equally evident that he does only a thousandth part of what he seems to do. Let us see, if we accept this theory, to what sort of conclusions it will lead us. One conclusion to which it will lead us at once is the following—that each of us is responsible only for a thousandth part of his actions; and from this will follow others more remarkable still. Since the holiest man has elements of evil in him, and the worst man elements of good, the good deeds for which we honour the saint may

really be the result of his antecedents, and his few
faulty deeds may be all that we are to attribute to himself ; whilst, conversely, the criminal's antecedents may have been the cause of all his crimes and vices, and he may himself have done nothing but some acts of unnoticed kindness. It will be thus impossible to form any true judgment of anybody ; for the real St. Peter may have been merely a false and truculent ruffian, and the real Judas Iscariot may have been fit for Abraham's bosom. And yet even these conclusions deducible from the premises of Mr. Bellamy are sane when compared with those deducible from the premises of Mr. Sidney Webb ; for Mr. Bellamy would allow a man to be responsible for a thousandth part of his actions at all events, whilst Mr. Sidney Webb would not allow that anybody either did or was responsible for anything.

And now, finally, let us turn to that other *Finally, let us take the argument which seeks to eliminate the causality of ment that most the great man, not by proving that he owes his of what the great man does superior brain-power to society, but by proving that depends on superior brain-power has little to do with his and achievements, their principal cause being the ap- he does but add a little.* pliances, the opportunities, and the accumulated knowledge at his command ; and that these, at all events, are due not to himself, but others—to the efforts of past generations, and the legacy they have left to the present. This is the argument which is mainly relied upon by Mr. Spencer. He insists on the fact that none of the great inventors or discoverers could have made their discoveries or inventions if centuries of past progress had not prepared the way for them. "*A Laplace, for instance,*" he says, "*could not have got very far with the* Mécanique Céleste *unless he had been aided by the slowly developed system of mathematics, which we trace back to its beginnings amongst the ancient Egyptians*"; and his many other illustrations are all of the same kind.

If ·we consider the meaning of this argument carefully we shall see that its logical outcome is not to deny to the great man all superiority whatsoever, but to exhibit his superiority as being less than it is usually supposed to be. Laplace, Mr. Spencer would say, may have been personally a little above the level of his contemporaries, but he owed most of his elevation to sitting on the shoulders of his predecessors. Now if this reduction of the great man's reputed greatness to such very small proportions has any practical meaning, it must mean that greatness is not only less than it is supposed to be, but is also a great deal commoner, and more easily procurable. Whatever any particular great man has done, could have been done, if he had not done it, by an indefinite number of others. Let us then take as an illustration some definite task, and see how far such reasoning has any practical application. Our illustration shall be taken from the domain of art; for the great artist, according to Mr. Spencer's special statement, owes his greatness to the achievements of past generations, just as the great mathematician does, or the great thinker, or the great

If this argu-
ment means
anything
practical, it
must mean
that greatness
is commoner
than it is
vulgarly
thought.

inventor. Let us suppose, then, that it is desired to decorate some public hall with pictures worthy of Titian or Michael Angelo, or to open some national theatre with a new play worthy of Shakespeare. The great question will be where to find the artist or poet whose works shall even approximate to these ideals of excellence; and any one who knows anything about either pictures or poetry will know that to find him is a well-nigh hopeless task. Now what conceivable help, what conceivable meaning, would there be in Mr. Spencer's coming forward and telling the public that the greatest poet or artist is the product of the same conditions that have produced any one of themselves ? Mr. Spencer has actually made this precise statement. Let us therefore refer to the terms in which he has done so. " *Given a Shakespeare,*" he says, "*and what dramas could he have written, without the multitudinous conditions of civilised life —without the various traditions which, descending to him from the past, gave wealth to his thought, and without the language which a hundred generations had developed and enriched by use ?* "

Mr. Spencer could not have put his own case more clearly ; and the more clearly it is put, the more easy it is to answer it, and to show that for practical men it has no meaning whatsoever. The answer to the question he asks is not only obvious, but contains at the same time the solution of the whole problem we are discussing. It will inevitably take the form of another question. Given the

Book I
Chapter 3

But is this the case? Does Shakespeare's debt to his antecedents make Shakespeares more numerous?

Book I
Chapter 3

conditions of civilised life, and the traditions of England and its language, as they were under Queen Elizabeth, how could these have produced dramas like *King Lear* and *Hamlet*, unless England had happened to possess that unique phenomenon — a Shakespeare? Could a Bottom have written these dramas, or a Dogberry, or a Sir Toby Belch? Or could Sir Thomas Lucy, or any of the "poetasters" satirised by Ben Jonson? Or could the actors, Kemp, Jones, and Bryan, who assisted in the representation of these dramas upon the stage? The answer is, of course, No. And yet these men

Shakespeare's contemporaries had the same national antecedents that he had; but they could not do what he did.

inherited the same language that Shakespeare did; the three last had the advantage of knowing his finest passages by heart. The weaver, the bellows-mender, the constable, the Justice of Peace, had behind them the same traditions that Shakespeare had, and were surrounded by the same "*multitudinous conditions*" of civilisation. But out of these conditions one man alone was capable of eliciting the results elicited by Shakespeare. The real explanation of the whole difficulty — the difficulty involved in the fact that whilst the argument of Mr. Spencer and Mr. Bellamy is, in a speculative sense, not merely true but a truism, it is utterly untrue in any practical sense—is as follows : Every human being living at any given time is, as Mr. Spencer says, an inheritor

Men inherit the past only in so far as they can assimilate it,

of the past ; but men inherit the past in very different degrees. They inherit the knowledge of the past only according to the degree to which they acquire it ; the language of the past only according to their skill in manipulating it; the inventions of the past only according to their skill in reproducing and using them.

The extraordinary confusion of thought in-volved in Mr. Spencer's position is focalised in an argument constantly employed by socialists—that *" inventions once made, become common property."* Except the earliest and simplest of them, they no more become common property than the count-less facts and figures buried in Parliamentary Blue-Books become the property of every new member of Parliament, or than encyclopædic knowledge becomes the property of every one who happens to inherit an edition of the *Encyclopædia Britannica ;* or than the power of deciphering the hieroglyphics which are preserved in the British Museum becomes the property of every cabman who drives his vehicle along Great Russell Street. It is perfectly true that the discovery of each new portion of knowledge enables men to acquire it who never might have discovered it for themselves ; but as the acquisition of the details of knowledge becomes facilitated, the number of details to be acquired increases at the same time ; and the increased ease of acquiring each ,is accompanied by an increased difficulty in acquiring all, or even in assimilating those which are practically connected with one another. A mechanic, for instance, could with ten minutes' attention comprehend the principle involved in a cantilever bridge, but to design and construct a bridge such as that which now spans the

6

Forth, with its spans of six hundred yards and its altitudes of aerial steel, implies an assimilation of our multitudinous existing knowledge, such as is hardly to be found in a score of engineers in Europe. Or to turn once more to Mr. Spencer's example of Shakespeare, whilst all Shakespeare's contemporaries had the same antecedents that he had, the same line of thinkers behind them, and the same developed vocabulary, Shakespeare's mind was capable of absorbing much of the national inheritance, whilst the great mass of his contemporaries could comparatively absorb very little.

The discoveries
and inventions
of the past are
the property
of those only
who can ab-
sorb and use
them.

We are thus brought back to the point from which we set out—namely, the differences in capacity by which men are distinguished from one another; and we see that all the reasonings of our modern sociologists have, for practical purposes, left these differences undiminished. The difference between the great man and the ordinary man is not made less by the fact that they both of them owe much to a common past, any more than the difference between a hogshead of water and a wine-glass is made less by the fact that both have been filled from the same stream.

Thus the intro-
duction of the
past into the
question
leaves the
differences be-
tween the great
man and
others un-
diminished.

The conclusion, therefore, of the whole matter is as follows. In the first place, whatever may be the speculative significance of Mr. Spencer's contention, which Mr. Bellamy expresses with the arithmetical precision of an accountant, that each living generation does only a minute fraction of what it seems to do, or of arguments like Mr. Sidney Webb's, that

each living generation does nothing at all of what
it seems to do, the mass of living men at all events do something, in the very real sense that if they did not do it they would die ; and the doing of this *If the ordinary man does anything, the great man does a great deal more ;* something is for them the whole of life, and all practical problems depend on the manner in which they do it. Such being the case, it follows, in the second place, that however much the ordinary man does, the great man does a great deal more. Therefore, if the ordinary man does any of the things that he seems to do, and causes any of the events that he seems to cause—if he ploughs the farm that he seems to plough, and lays the bricks that he seems to lay—indeed we may add, if he eats the dinners that he seems to eat—the great man in a precisely similar sense is the cause of those changes and of that progress which he seems to cause. Hence of these changes he is, for the practical sociologist, not merely the proximate initiator, whose action and *and in practical reasoning he is a true cause for the sociologist.* peculiarities may be neglected, but a true and primary cause, on which the attention of the socio- logist must be concentrated ; and just as in action it is impossible to do without him, so in practical reasoning it is impossible to go behind him.

The reader has now been shown the absolute futility of that train of reasoning by which even so keen a thinker as Mr. Spencer has persuaded him- self that he can get rid of the causality of the great man, and in which every socialistic reformer who has risen above the level of a demagogue has attempted to find a scientific foundation for his im-

possible castle in the air.   But the demolition and
exposure of these mischievous and miserable fallacies
shall not be entrusted only to the arguments that
have been brought to bear on them.   The validity
of these arguments shall now be finally substantiated
by direct appeal to a sociologist whose identity may
surprise the reader.   This is none other than Mr.
Spencer himself, who, when he forgets to be the
conscious expositor of his theory, and turns aside
to illustrate some particular point by examples
drawn from the experience of common life, is constantly contradicting, in a most remarkable but
entirely unconscious way, the fundamental principle
which he has deliberately set himself to establish.

In the seventh chapter of his *Study of Sociology*,
being incidentally concerned to insist on the iniquity
and the mischievousness of war, he describes how
Europe, during the earlier years of this century,
was visited by certain disasters, far-reaching and
horrible, from the consequence of which the world has
hardly yet recovered.   These disasters consisted of
slaughter, plunder, pestilence, agony, rape, and ruin ;
and to say nothing of their results on those whom
they left alive, they resulted in some two million
He declares
that the Napo-
leonic wars
were entirely
due to the
maleficent
greatness of
Napoleon. violent and unnecessary deaths.   And how does Mr.
Spencer explain these appalling phenomena ?   He
who declares that we should learn nothing about
social causation "*should we read ourselves blind over
the biographies*" of all the great rulers of the world,
explains them by tracing them to one sole and
single cause ; and this, he says, was ·the genius

and personality of Napoleon. "*Out of the sanguin-ary chaos of the Revolution,*" he writes, "*rose a soldier whose immense ability, joined with his absolute unscrupulousness, made him now general, now consul, now autocrat. The instincts of the savage were scarcely at all qualified in him by what we call moral sentiments. . . . And all this slaughter, all this suffering, all this devastation was gone through—*" Let us pause and ask why it was gone through, according to Mr. Spencer. Does he say it was gone through because of "*aggregates of past conditions*" and the influence of antecedent generations? Far from it. He says, "*All this was gone through because one man had a restless desire to be despot over all men.*"

But perhaps Mr. Spencer may have a defence ready. He may tell us that the influence of Napoleon was merely that of a military leader, which influence he has excepted from his theory of general causes. To this it must be answered in the first place that Napoleon was at all events not a leader in "*early*" or "*primitive*" warfare, to which Mr. Spencer's exception is specifically and emphatically limited. Mr. Spencer consequently shows us, by his own practical reasoning, that this theoretical limitation of which he made so much cannot be maintained for a moment, and that what-ever is true of great leaders in a primitive war, he himself recognises—all his theories notwithstanding —as equally true of them in the most advanced stages of civilisation. But a far more important answer, and one taken from himself, is still in reserve — an answer which clenches the whole matter, and shows us that Mr. Spencer, in his dealings with practical life, really recognises great men as exercising in the arts of peace precisely the same kind of causality which Napoleon exercised in war.

Let us turn to Mr. Spencer's treatise on *Social Statics*, and to the section of it in which he treats of patents—or as he himself describes them, "*the rights of property in ideas.*" He begins by complaining that the right of patenting "*inventions, patterns, or designs*" is not recognised as being based on any moral right at all, but is generally regarded as a kind of "*privilege*" or "*reward.*"

He defends patents because they represent the very substance of the inventor's own mind;

"*The prevalence of such a belief,*" says Mr. Spencer, "*is by no means creditable to the national conscience. . . . To think,*" he exclaims, "*that a sinecurist should be held to have a 'vested interest' in his office, and a just title to compensation if it is abolished; and yet that an invention over which no end of mental toil has been spent, and on which the poor mechanic has laid out perhaps his last sixpence—an invention which he has completed entirely by his own labour and with his own materials—has wrought, as it were, out of the very substance of his own mind—should not be acknowledged as his property!*"

Social Statics is one of Mr. Spencer's earlier works. Let us now consult his latest, the third and final volume of his *Principles of Sociology;* and here we shall find this same admission that the

great man's achievements are wrought not out of
aggregates of conditions, but "*out of the very
substance of his own mind*," emphasised by him
as a practical truth, with all the vigour of a practical
man. In his chapter on the "*Interdependence and
Integration of Industrial Institutions*" he dwells with and he attributes the
modern im-
provement in
steel manu-
facture to Sir
H. Bessemer.
much eloquence on the almost incalculable benefits
that have resulted, and extended themselves through
the whole industrial world, from certain improvements introduced into the manufacture of steel.
And to what were these improvements due ? Mr.
Spencer answers this question not merely by admitting, but by insisting with the fervour of a
hero-worshipper, that they were due to the genius
of one single man, namely Bessemer ; and so obvious
does this truth appear to him, that he devotes an
indignant footnote to denouncing the governing
classes for not being sufficiently alive to it, and for
conferring on a man who, "*out of the very substance
of his own mind*," had wrought such gigantic and
universally beneficial changes, no higher reward
than the title of Sir Henry Bessemer—"*an honour*"
he says, "*like that accorded to a third-rate public
official on his retirement, or to a provincial mayor on
the occasion of the Queen's Jubilee.*"

After this, what more need be said ? Here we
have Mr. Spencer himself, the moment he touches
the practical side of life, contemptuously brushing
aside the whole of his speculative theory and admitting, or rather insisting, with the most unhesitating
and uncompromising vigour, that "*the phenomena of*

social evolution," even if they do not result entirely, as Carlyle would have it, from the actions of great men, yet cannot, at all events, be possibly explained without them; and that great men, their natures, and the details of their active lives, are primary factors to be studied by every practical sociologist, and are not to be merged in "*society,*" in "*antecedents,*" and in "*aggregates of conditions.*"

So much, then, being established, we must consider two difficulties suggested by it. The practically independent character of the great man's causality will be yet more apparent at another stage of our argument, and we shall see that the whole structure of all civilised societies depends on it. But we may, for the present, regard it as being sufficiently established, and the absurd and unreal character of the attempts to get rid of it demonstrated. So much, then, being assumed, we will, in the following chapter, consider two objections of a character very different from any of those of which we have now disposed. They are objections which will very possibly have suggested themselves to the reader's mind, but which, instead of conflicting with the truth which has been just elucidated, will be found, when examined carefully, to emphasise and to enlarge its significance.

CHAPTER IV

THE GREAT MAN AS DISTINGUISHED FROM THE
PHYSIOLOGICALLY FITTEST SURVIVOR

THE two objections to which reference has just been It may be objected that modern sociology does not, as here asserted, neglect the great man, for it adopts the doctrine of the survival of the fittest. made are connected with two doctrines, neither of which has thus far been submitted to any detailed examination, and one of which has indeed been hardly so much as alluded to, but which are both intimately associated, in the estimation of the world at large, with contemporary science, and more especially with contemporary sociology. One of these doctrines is that of the survival of the fittest. The other is that which, more or less distinctly, is suggested at the present time by the much-abused word "evolution." When the reader thinks of the doctrine of the survival of the fittest, when he reflects on the fact that Mr. Spencer is an avowed disciple of Darwin, and that Mr. Spencer's own disciples are constantly making allusion to "*the rivalry of existence*" and the "*successfuls and the unsuccessfuls*," he may be tempted to ask himself if it can be really true that Mr. Spencer has eliminated the great man from his system after all. On the other hand, when the reader thinks of *evolution*,

Book I
Chapter 4

which, whatever it may mean, at all events means a progress essentially different from the achievements of particular individuals, he may wonder in what way the doctrine of evolution can be reconciled with any doctrine which has the achievements of individuals for its basis.

It may be asked, on the other hand, what place the great man has in an exclusively evolutionary theory of progress.

We will take these two points in order. With regard to the survival of the fittest in the competitive struggle for existence, the great fact which it is necessary to make clear is as follows ; and it is one which our contemporary sociologists have altogether failed to perceive. In the evolution of societies, just as in the evolution of species—in the evolution of man as a social being, as in the evolution of man as an animal—the struggle for existence has played an important part ; but in social evolution the part played by it is very far from being that which is popularly supposed, nor does the survival of the fittest in any way correspond with the position and influence claimed for the great man. Certain of the phenomena of progress are no doubt produced by it, but they are as different from those which the great or exceptional man produces, as is the movement of the earth round the sun from its movement round its own axis. In order to understand this, let us first consider carefully how progress, as the result of the struggle for existence, is explained by our contemporary sociologists. The matter is put succinctly and very clearly in the following passage from Mr. Kidd's *Social Evolution.*

The fittest survivor is not the same as the great man.

He plays a part in progress, but not the same part.

" *Progress everywhere,*" he says, " *from the begin-*

ning of life, has been effected in the same way, and is
possible in no other way. It is the result of selection
and rejection. In the human species, as in every
other species which has ever existed, no two indi- The fittest
viduals of a generation are alike in all respects ; there men, by sur-
viving, raise
is infinite variation within certain narrow limits. the general
level of the
Some are slightly above the average in a particular race, and pro-
direction, as others are slightly below it ; and it is mote progress
only in this .
only when the conditions prevail that are favour- way.
able to the preponderating reproduction of the former,
that advance in any direction becomes possible. To
formulate this as the immutable law of progress since
the beginning of life has been one of the principal
results of the biological science of the nineteenth
century. . . . To put it · in words used by Professor
Flower in speaking of human society, 'Progress has
been due to the opportunity of those individuals who
are a little superior in some respects to their fellows
of asserting their superiority, and of continuing to
live, and of promulgating as an inheritance that
superiority.'"

The entire Spencerian position as regards the
social struggle for existence is here given us in a
nutshell. The competitive struggle is a process
which produces progress by means of the manner
in which it affects men in general. In any community the means of subsistence are being constantly
appropriated by the members who are a little
stronger than the rest, whilst those who are weaker
have an insufficient portion left them. The latter
therefore die early themselves; or breed no children;

or breed children who die early ; whilst the former live long, and breed children who live likewise ; and of these children there is always a certain percentage in whom are reproduced the superior qualities of their parents. Thus the weaker members of the community are always dying out, whilst stronger members not only become more numerous, but more efficient as individuals also. In other words, the Darwinian struggle for existence produces progress by raising the general average of efficiency. It has nothing to do with a few men towering over the rest. It works by producing a simultaneous rise of all. The superior "*assert their authority*" not by commanding their inferiors, but merely by "*continuing to live*" and having children as superior as themselves. In this way, to quote an illustration from Mr. Spencer, the progressive races of Europe have reached a stage of development which makes possible amongst them the appearance of men like Laplace or Newton, an event which could not occur amongst the Hottentots or the Andaman islanders. It will thus at once be clear that the theory of the survival of the fittest explains progress by reference

The great man promotes progress by being superior to his contemporaries.
to an order of facts totally distinct from those involved in the influence claimed for the great man. Whilst the theory of survival is illustrated by the superiority of Europeans to Hottentots, the greatman theory is illustrated by, and depends on, the superiority of men like Newton to the great mass of Europeans.

What relation, then, do these two explanations

bear to each other ? In a direct way they are not related at all. They neither conflict with each other nor overlap each other. They are both of them true ; but true as explaining different sets of pheno- mena. One of the great errors of which our modern sociologists are guilty consists in their failure to perceive that social progress is not a single move- ment but the joint result of two, which differ from each other—to repeat what was said just now—quite as much as do the two movements of the earth. The difference between them will become instantly clear to us if we will turn our attention merely to the single obvious fact that the two take place at different rates of speed, the one set of changes being slow, like the succession of years ; the other set of changes being rapid, like the succession of days. The general rise in capacity which distinguishes the modern civilised nations from primitive man, or from the lowest savages of to-day, and which has been due to what Mr. Kidd calls "*the preponder-* *ating reproduction of individuals slightly above the average,*" has been the work of an incalculable number of centuries. It has been so slow that, in many respects at all events, it has been indistin- guishable during the course of several thousand years. The great thinkers amongst the ancient Egyptians were not congenitally inferior to the great thinkers of to-day. The brain of Aristotle was equal to the brain of Newton ; whilst the masons whose hands constructed the Coliseum and the Parthenon knew as much of their craft as those who constructed the Imperial Institute. But with this slowness in the rise of the general level of capacity let us compare the progressive results achieved within some short period. We cannot do better than take the past hundred years, and consider the progress made in the material arts of life. How the whole spectacle changes! Within that short period, at all events, no one will venture to maintain that the average congenital capacities of our own countrymen have been enlarged. We are not wittier than Horace Walpole, more polite than Lord Chesterfield, more shrewd and sensible than Dr. Johnson ; whilst it is easy to see by reference to those trades, such as the building trade, which science and invention have done comparatively little to alter, that the natural efficiency of the average workman is no greater now than in the days of our great-great-grandfathers. And yet during that short period what an astounding progress has taken place! To sum it up in a bald economic formula, whilst the capacities of the average Englishman have remained altogether stationary, the economic productivity per head of the population of this country has during the past century trebled, and more than trebled itself.

This remarkable comparison between the rapidity of actual progress and the extreme slowness of the biological development resulting from the survival of the fittest in the Darwinian struggle for existence, will be enough to show anybody that progress is not one movement but two ; and whilst the survival of the fittest explains the slow and almost imperceptible movement, the rapid and perceptible movement is explained by the leadership of the greatest. It is with the rapid movement alone that the practical sociologist is concerned ; and hence for him the great man, not the fittest, is the important factor.

Let us now consider what is meant by the process called social evolution, regarded as something distinct from those intentional advances made and maintained by the genius of great men. To understand this, we must consider what is meant by evolution generally. Mr. Spencer defines it in terms of "*the homogenous*" and "*the heterogenous*"; and from his own point of view we may accept his definition as correct. But facts have many aspects ; and according to the purpose with which we deal with them they will require different definitions, which, though none of them are incompatible with the others, will have between themselves no apparent resemblance. Thus the biologist's definition of a man will be quite distinct from the theologian's ; and the dangerous illness of a great party leader will be one phenomenon for his followers, and quite another for his doctor. In the same way Mr. Spencer's definition of evolution, however admirable it may be from a certain point of view, is not exhaustive. It entirely leaves out of sight those characteristics of the process which it is necessary before all things that the practical sociologist should understand.

To reach a definition that will include these

Book I
Chapter 4

Its great prac-
tical character-
istic, as put
forward by
Darwin, is that
it is opposed to
the doctrine of
design, or
divine inten-
tion ;
let us begin by fixing our attention on that order of facts which formed the special study of Darwin, and in connection with which the theory of evolution became first known to the world; and let us ask what was the greatest and the most notorious effect produced by Darwinism on human thought generally. Its greatest and most notorious effect was to disprove, or rather render superfluous, the old theory which explained the varieties of organic life by referring them to the design of some quasi-human intelligence. According to the old theory, every species of living thing, from the lowest to the highest, was produced by the power and purpose of one supreme Mind, who adapted the frame and faculties of each to a pre-arranged set of circumstances and the fulfilment of certain needs. According to the theory of evolution, associated with the name of Darwin, these results were accomplished by purpose and intelligent power likewise, only not by the power and purpose of one supreme external Mind, but by the power and and yet, according to Darwin,
species resulted
from the inten-
tion of each
animal to live
and propagate. purpose of the living things themselves. Each living thing chose its mates, reproduced its kind, hunted for food, fought with rivals, and either conquered or was conquered by them, in obedience to the promptings of its own instinctive purposes. These were the motive power of the whole evolutionary process. The variety and development of organic life, as we know it, did not result indeed from one great intention, but it did result from an infinity of little intentions.

Now so far the theory of design and the theory of evolution very closely resemble each other ; but here we come to the point of essential difference between them. According to the theory of design, the varieties and gradations of organic life were not only the result of intention in the supreme Mind, but were also themselves the exact result intended. According to the evolutionary theory, although they were the result of an infinity of intentions, not one of the living things, from whose intention they resulted, intended them. They were the by-product of actions undertaken for entirely different ends—that is to say, for the benefit of the individual creatures who undertook them. This is the essential and this is the peculiar character with which the theory of evolution invested them. It presented to the mind the extraordinary phenomenon of a single series of actions producing a double series of results—the intended and the unintended, the latter of which, though entirely different from the former, was equally orderly, equally reasonable and coherent. Evolution, in fact, as revealed to us in the physiological world, is, for the philosopher, neither more nor less than this—the *reasonable sequence of the unintended*.

But this definition of evolution does not apply only to development in that world of facts studied by Darwinian science. It is equally applicable to the process of social evolution also. Indeed social evolution is even more strikingly, though not more truly, than physiological evolution, the reasonable

Book 1
Chapter 4

Species, therefore, according to the evolutionist, is the result of intention, but not the result intended.

Evolution, in fact, is the reasonable sequence of the unintended.

This is as true of social evolution as it is of biological.

7

sequence of the unintended. How this is can be easily made plain ; and when once the idea is grasped, which the definition embodies, it will be seen that social evolution, although it is no doubt different from all or from any of those changes deliberately produced by the agency of the great man, instead of excluding these changes, or eliminating the great man as the cause of them, is a process which depends altogether upon him and them, and that, instead of obscuring the great man's importance, it only exhibits it in a stronger and clearer light.

Many of the social conditions of any age result from the past, but were intended by nobody in the past ;

Let us take then our definition of evolution as the reasonable sequence of the unintended, and apply the idea embodied in it to that aggregate of conditions, either in our own or any similar period, amongst which the great man works. All these conditions, such as the knowledge which he finds accumulated, the inventions which he finds in use, the political and the economic conditions of his country, are, taken as a whole, the result of no one man's genius. It is equally obvious that they do not, in their incalculably complex entirety, represent any one man's intention, or even the joint intention of any number of men acting in concert. Printing,

for instance, many of the social effects of railways and cheap printing.

for example, and railway travelling have produced a number of social results never dreamed of by the men who perfected our locomotives and our steam printing presses. Accordingly, when any great man of to-day initiates some fresh social change, whether as an inventor, a director of industry, a politician, or a religious teacher, a large part of his achievement consists in his manipulation and refashioning of results of past human action, which can be set down to the credit, or ascribed to the intentions of no individual, and no body of individuals. The society of the past intended these no more than the great men of the past. They are results, that is to say, which come all under the heading of the unintended. But when we consider the great man's achievement thus, we shall not only witness the grouping of many of the factors essential to it into one heterogeneous but logically coherent class, as the unintended. When such a grouping has taken place, we shall see that there remains behind an equally coherent and equally striking residuum—namely, the social results and conditions that have been obviously and notoriously intended. These may not be found existing apart from the former; but though in conjunction or combination with them, they will be visible as a distinct and separate element, and their true importance as a factor in social progress will begin to be apparent to the mind the moment their specific peculiarity, as just described, is apprehended.

Let us take a few examples which, owing to their magnitude and familiarity, will be at once intelligible. Our first shall be taken from the histories of art and of speculative philosophy. In each of these domains of human activity and achievement we find those phenomena of development to which it is now customary to apply the name of evolution. Thus we hear of the evolution of philosophy from the

Book 1
Chapter 4

Therefore, whenever any great man produces some change intentionally he has to work with unintended materials.

We can see this in the progress of dramatic art ;

crude guesses of Thales to the elaborate system of Aristotle. We hear of the evolution of the Greek drama from the exhibitions of Thespis with his cart to the tragedies of Æschylus and of Sophocles; and similarly we hear of the evolution of the English drama from such exhibitions as miracle plays or *Gammer Gurton's Needle* to tragedies such as *Hamlet* and comedies such as *As You Like It.* And to all such examples of development the word *evolution* is applied with perfect accuracy; for there is in each an obvious and orderly sequence of the

unintended. Aristotle's philosophy was in part derived from that of his predecessors. He employed existing materials so as to produce a result which was not intended, indeed was not even imagined, by those who originally got them together and fashioned them, and which would never have been reached by Aristotle himself, if his predecessors had not thus unintentionally assisted him. None the less, however, does the Aristotelian philosophy, as its author gave it to the world, embody the deliberate intention of his profound and unrivalled genius; and it is only because it embodies this intended element that it constitutes an advance on the philosophies that went before it. Similarly, though

And yet in
each case the
intended
elements equal
or are greater
than the
unintended. Sophocles and Shakespeare, in constructing their dramas, each profited by the achievements of the dramatists who had gone before them, and though the art of each would doubtless have been more crude and imperfect had he come into the world a generation or two before he did, yet the part played

by evolution in the production of *Hamlet* and
Antigone is totally distinct from, and is altogether
dwarfed by, the part played by the genius and the
intentions of their great authors.

Let us now turn to invention and applied science ; We see the
same thing in
the history of
the *Times*
printing press.
and the history of social progress, as connected
with and derived from them, will show the same
two elements — the unintended and the intended,
similarly related and similarly coexistent. A
brilliant illustration of this fact is provided for
us, in one of his books, by Mr. Herbert Spencer,
though he himself, with a curious blindness and
perversity, uses it not to illustrate but to obscure the point on which we are now dwelling.
The illustration referred to is the history of the
press by which the *Times* is printed, which implement, according to Mr. Spencer, is the result
altogether of evolution. *"In the first place,"* he
says, *"this automatic printing machine is lineally
descended from other automatic printing machines
. . . each pre-supposing others that went before. . . .
And then, in tracing the more remote antecedents, we
find an ancestry of hand printing presses."* He
further points out that this press implies not only an
ancestry of former presses, but also the existence of
the machinery used in making it, and again how this
machine-making machinery has a distinct ancestry
of its own, which includes the fact of the abundance
of iron in England. Geometry, physics, chemistry
also, he says, played their part in the process ; and
he winds up by referring to purely social causes.

Why, he asks, was the Walter press produced?   In order that "*with great promptness*" it might "*meet an enormous demand.*"

It is difficult to imagine a better illustration than this of the part played by evolution in the domain of mechanical invention.   It is perfectly evident that the mass of discoveries and inventions which preceded and paved the way for the final invention in question were due to men who had no idea in their heads of such a machine as a steam-driven printing press at all.   When printing was first invented, steam-power was undreamed of.   When the steam-engine was being perfected as a means of driving machinery, the inventors had no specific intention of applying this force to the printing press. The men whose genius and energy in the seventeenth and eighteenth centuries laid the foundation of the English iron trade, and with it, as Mr. Spencer says, the foundation of "*machine-making generally,*" in all probability never even saw a newspaper, and could not have conceived the possibility of collecting enough news daily to fill as much as one page of the *Times.*   The mathematicians and chemists to whose work Mr. Spencer alludes most probably never gave a thought to the practical application of their discoveries, and knew as little of the process of printing as they did of Chinese grammar. But let us give to these facts all the weight we can. Let us accept the antecedents that made the Walter press possible as not only sequences but also concurrences of the unintended ; and yet the part played

It was the
result of many
kinds of un-
intended
progress, con-
stantly re-
combined by
intention.

by the great man remains as essential, and remains as large as ever. The fact that the Walter press could never have existed unless Caxton's press had existed, and that Caxton never foresaw the future development of his apparatus, does nothing to disprove the fact that in the development of printing generally, genius like Caxton's was an indispensable agent, and one which stamped its character on the whole sequence of inventions which it inaugurated. Nor again does the fact that an invention like the Walter press implies not only a direct sequence of inventions and discoveries, but also a concurrence of many separate sequences, such as the invention and discoveries of chemists, of machine-makers, and producers of iron, do anything to disprove the importance and the necessity of the part played by the men to whose genius the press was directly due. For although the co-existence of the separate sequences referred to —the parallel march of progress in many separate arts and sciences—may have been altogether un-intended by any of those concerned in them, what was emphatically not unintended was their final concurrence—the fact of their being brought together for one definite purpose. This was due to the deliberate intention of exceptional men with strong synthetic powers, who appropriated and connected the achievements of various other men. Chemistry, geometry, the production of iron, and the develop-ment of machinery for machine-making would never have worked together to produce an automatic

printing press had the immediate inventors of such
an implement not coerced them into their service,
and forced them to contribute to a deliberately
planned result.

Evolution, in
fact, is the
unintended
result of the
intentions of
great men.
The state of the case is this.   Let us take any
civilised society at any period we will, and examine
it in the act of advancing to the next stage of its
development.   We shall find that its existing conditions consist partly of results intended by particular
great men whose past actions have produced them,
and partly of results neither foreseen nor intended
by anybody.   Thus at the present day amongst
our social conditions we have the telegraph and the
railway system—both of them results intentionally
produced by individuals ; and we have also a
variety of new wants and habits, new methods of
conducting trade and government, which have been
produced by these, but which were neither intended
nor even thought of by the inventors of the locomotive, or by Wheatstone and Cooke when their
wires at last realised the long-forgotten dream of
the Italian Jesuit Strada.[1]   Thus, though social
conditions at any given time are a compound of
intended results and unintended, and even though
we may admit that at any given time these last are
more widely diffused than the former, these last

[1] Strada, an Italian Jesuit, in the seventeenth century invented,
or rather imagined, communication by electric telegraph ; and his
idea actually comprised the use of two needles moved by two magnets,
these magnets being connected in such a way that, by the movement of either of them, the needle actuated by the other could be
made to point to such and such letters on a dial.

are themselves the children of intention once removed. Great men may not have meant to
produce them, but they have arisen from conditions
which great men did mean to produce; and they
could not have arisen in any other way. And here
we are brought to a fact more obvious and more
important still. Before any further advance in social
civilisation can be made, other existing conditions,
whether intentionally produced or not, require to be
intentionally re-combined and acted on by men whose
enterprise, whose intellect, and whose constructive
imagination mark them out from their fellows as
the pioneers of the future. We are thus once
more confronted with the fact already insisted on—
that the social conditions of a time are the same
for all, but that it is only exceptional men who can
make exceptional use of them, and turn them into a
stepping-stone on which their generation may rise
higher.

Social evolution, therefore, in so far as it
is other than biological, may be defined as the
unintended result of the intentions of great men;
and this definition at once brings us back to the
truth which was urged in the first chapter as the
starting-point of our argument, and which can now
be put before the reader with an added force and
clearness.

It was said in the first chapter that sociologists
have succeeded in dealing with those wider social
phenomena which are exhibited by social aggregates
as wholes, and which are interesting and significant

to the speculative or religious philosopher. The truth of this statement is illustrated by what has what concerns the speculative philosopher. just been said about evolution. If evolved phenomena are phenomena which exhibit a reasonable sequence, and have yet been intended by no animal or human mind, it is open to the thinker to argue that they must have been intended by the mind of some higher power; and a new gate is opened into the Eden of theological speculation, from which man was driven when he first ate of the tree of scientific knowledge.

The intended element, which originates directly in the great man, is what is of interest for practical purposes. But whilst the business of the speculative philosopher is solely with the phenomena that have been unintended, the business of the practical sociologist is solely with the phenomena that have been intended. A moment's reflection will convince the reader that this must be so. The meaning of the words "practical science" is a science from which we can draw practical advice; but all advice implies an intended end; and every attempt to solve social problems scientifically must be concerned with results which we may deliberately set ourselves to produce, and not with by-products which, *ex hypothesi*, are beyond our calculation. We may study these by-products of intention as they have shown themselves in the past; but if we do this, it will be with the object of becoming able to foresee them in the future. So soon as we can foresee them, we shall be able to intend their production; and when this happens they will cease to belong to the unintended. The great man will then consciously aim at them, and

not leave them to the incalculable chances of evolution. It may safely be said, no doubt, that,
let us study human conduct as we may, unintended, or evolved phenomena will always continue to form a large part of what we mean by social progress ; but, as practical inquirers, we must put them on one side, and confine our attention to those factors in the problem which either embody some definite human intention themselves, or on which we can found, by studying them, some definite intention of our own. And of such factors the chief is the great man, whose importance is enhanced rather than dwarfed by the fact that his intellect and his energy are the causes not only of great results which he intends, but also of those others—wider, if not more important—which, though neither intended nor fore-seen by himself or by anybody else, would, if it were not for him, never take place at all.

BOOK II

CHAPTER I

THE NATURE AND DEGREES OF THE SUPERIORITIES
OF GREAT MEN

THAT great men are true causes of progress is The causality of the great man being established, we must consider more precisely what greatness is. admitted by Mr. Spencer himself to be the natural opinion of mankind. What has been done, then, in the preceding book is not much more than this :—a sound popular judgment, which is of the highest sociological importance, has been rescued from the discredit cast on it by the sophisms of modern theorists. These very theorists themselves, when they reason as practical men, have been shown to the reader blowing all their disproofs of it to the winds, and holding and appealing to it as tenaciously and as passionately as anybody ; and it is consequently given back to us, with its old authority unimpaired. Sound popular judgments, however, are not science. They lack what is the essence of science—that is to say, analytical precision. We must now, therefore, take this judgment with regard to the great man, and endeavour to invest it with a meaning exact and full enough to enable us to apply it to the detailed phenomena of society.

And here Mr. Herbert Spencer shall once more

help us ; for this remarkable writer, though he fails
to recognise what he is doing, not only appeals on
many critical occasions to the great - man theory
as an explanation of the most important social
phenomena, but he is repeatedly calling attention
throughout his sociological writings to those facts of
human nature of which the great-man theory is the
expression.   It will be sufficient to quote a few
passages only.

Mr. Spencer
will help us to
a general
definition of it. Let us turn, then, to the opening pages of Mr.
Spencer's *Study of Sociology* and consider what is
contained in them.   We shall find that they are
entirely devoted to describing the abject mental
condition of by far the largest portion of all classes
of English society, from the labourer, the farmer,
and the Nonconformist minister with his Bible, up
to "men called educated" and the most illustrious
of our historians and philosophers.   All of them,
says Mr. Spencer, "*are slaves to unwarranted
opinions*"; "*proximate causes*" are all that the
majority of them are able to understand.   Nor does
he represent this as some accidental result, due to
prejudices or deficiencies in education peculiar to
our own country.   He represents it as an inevitable
result of the character of the human race.   In his
"*Postscript*" to the same volume he takes care to
make his meaning plain.   "*Most people*," he says,
"*conclude quickly from small evidence*," and are
incapable "*of comprehending in their totality
assembled propositions.*"   Indeed, those whose
mental constitution is such that they can take a

rational view of "*human affairs*" are, he proceeds
to say, merely "*a scattered few.*" He elsewhere divides society into "*the capable and the incapable,*" the "*worthy and the unworthy*"; and in the "*Postscript*" just alluded to he mentions as an admitted fact that in every social aggregate "*the inferior form the majority.*" But a yet more caustic passage remains to be mentioned. In this same work, *The Study of Sociology*, he is ridiculing—and very justly—the socialistic idea that the State can be endowed with any talent or wisdom beyond what happens to be possessed by the individual functionaries who compose the State. These functionaries, he says, are merely "*a cluster of men,*" which, like any other cluster taken at hap-hazard, will comprise "*a. few clever individuals, many ordinary, some decidedly stupid*"; and he devotes pages to showing by means of multiplied examples, how incapable the ordinary statesman, to say nothing of the decidedly stupid, has been of promoting progress in even the simplest ways.

Mankind at large, then, according to Mr. Spencer, may, roughly speaking, be divided into three classes — the "*clever*" who are few, the "*ordinary*" who are the bulk of the population, and the "*decidedly stupid*" who form a considerable residuum; and it will appear from what he says of that representative "*cluster,*" the State, that whilst all real progress is the work of the clever few, the "*ordinary men*" do nothing to promote it, and "*the decidedly stupid men*" impede it.

8

Now it must be perfectly obvious to the reader that in this description of mankind we have the fundamental facts before us which the great-man theory formulates. For let us begin by supposing that the entire human race contained no individuals superior to the "*decidedly stupid,*" who, whenever they are placed in official positions, do nothing, Mr. Spencer declares, but commit the most pernicious blunders, either by their irrational conservatism, or their still more irrational innovations. It is obvious that in this case the world would never have progressed at all. Let us next suppose that in addition to the "*decidedly stupid*" men, the human race comprises also a large proportion of "*ordinary*" men, but not a single man who deserves to be called more than "*ordinary.*" Could social progress, as we know it, have taken place even then? Could thought, for example, ever have made any advances, had everybody been as incapable as Mr. Spencer's "*ordinary*" man is of taking a rational view of human affairs — had everybody been enslaved, like him, "*to unwarranted opinions,*" and been, like him, entirely lacking in the faculty which enables a man to comprehend "*assembled propositions in their totality*"? Or to put the whole matter in terms of a single instance, could Mr. Spencer's own system of philosophy have been written if he himself had not been immensely superior not only to "*ordinary*" men, but even to those rival thinkers whom, in every one of his volumes, he treats with such supreme disdain?

The answer of course is No. Under such conditions
progress would have been quite impossible. Our simple argument will accordingly run thus. It is evident that those triumphs of thought, enterprise, and invention, to which social progress is due, could never have been made had the whole of each generation been as stupid and void of character as its lowest and weakest members. Therefore progress must be due to men who are superior to the "*decidedly stupid.*" Here we have the great-man theory in embryo. But it is equally evident that we can go a step farther, and say that progress could never have taken place had there been no individuals who in will, originality, and intellect were therefore pro-gress must be superior to "*ordinary men.*" Social progress, therefore, must be due to this third class—the class which clever, who are, as Mr. alone is capable of taking "*a rational*" view of Spencer says, a scattered few. things ; but this class, as Mr. Spencer tells us, consists of a "*scattered few,*" and here we have, in Mr. Spencer's own language, neither more nor less than the great-man theory developed. We have it developed in the form of a distinct general proposition that progress is due not to mankind at large, but to a minority of exceptional individuals, and in this form, which Mr. Spencer has assisted us in This is the great-man giving it, it is brought into ·actual accordance with theory reason-the facts of social life, and, unlike the wild exaggera- ably stated. tions of Carlyle, it will be found to accord the more closely with them the more fully it is analysed.

The error of writers like Carlyle was that they took a part for the whole. They recognised no.

great men at all except great men of the greatest kind—heroic figures which appeared once or twice in a century; and as for the rest of mankind, they

treated them, in accordance with Mr. Spencer's formula, as a mass of units, approximately equal in capacity. The truth of the case is, on the contrary, this :—that whatever is done by great men of the heroic type, something similar, if not so striking, is done by a number of lesser great men also; that whilst the action of the heroic great men is intermittent, the action of the lesser great men is constant ; and that the latter, as a body, although not individually, do incalculably more to promote progress than the former.

Let us accordingly make it perfectly clear that when we describe great men as being a minority, or

a "*scattered few,*" we do not mean that out of every thousand men there are nine hundred and ninety-nine "*ordinary*" men and one genius ; or that there are (let us say) seven hundred who can be described for all purposes as "*ordinary,*" and two hundred and ninety-nine who can be for all purposes described as "*stupid*"; and that there is one "*clever*" or "*great*" man who towers over them like an oak tree over bramble bushes. Nor, again, do we mean that "*greatness*" is some single definite quality, which marks its possessor out like a white man amongst negroes. Believers in extreme democracy, who very rightly discern in the great-man theory ᵥ the destruction of their favourite enthusiasms, will instinctively seek to attribute some meaning such as this to its exponents. But the great-man

theory, when properly analysed and explained, will
be found to comprise no such absurdities as the
foregoing. When we speak of "greatness" we Greatness is
mean a great variety of efficiencies, which, though various both in kind and
grouped together because they are all exceptional in degree,
degree, are nevertheless indefinitely various in kind ;
and, moreover, the degrees to which they are exceptional are indefinitely various also, the degree
being in many cases so low that it is difficult to say
whether it should be classed as exceptional at all.
In short, there are as many degrees of greatness as
there are of temperature ; and it is as difficult to
draw a line between ordinary men and men whose
greatness is of a very low degree, as it is to draw a
line between coldness, coolness, and low degrees
of heat. But though it may be questionable
whether we should call a day cool when the
thermometer is at fifty - nine, and whether we
should call it hot when the thermometer is at sixty-
one, everybody admits that it is hot when the but, at all
thermometer is at eighty-five, and cold when the a certain
thermometer registers twenty degrees of frost. In minority of
the same way, though there will be a certain number semble each
of people who may be classed as great by one judge more efficient
and classed as ordinary by another, there is a majority.
certain number whose capacities, however unequal
amongst themselves, set their possessors apart as
indubitably greater than the majority ; and we are
speaking with sufficient, though we cannot speak
with absolute precision, when we say that progress
depends on the action of this minority.

How great the inequality is between the natural powers of men is perhaps most clearly evidenced by the case of art, and more especially the art of poetry. In certain domains of effort it may be urged that unequal results are caused by unequal circumstances, quite as much as by unequal capacities. But about poetry, at all events, this cannot be said. Some of the greatest poets the world has ever known—it is enough to instance the cases of Burns and Shakespeare— have been men of no wealth and of very imperfect education. Obviously, therefore, in poetry one man has as good a chance as another. It is no doubt often argued—and this argument has already been examined—that great poets, of whom Shakespeare is a favourite example, owe part of their greatness not to themselves, but to their age. But this does nothing to explain the differences between poets who belong to the same age, and who, all of them, in this respect, start with the same advantage. Let us confine our comparisons then to men who were each other's contemporaries, and ask what made Burns a better poet than Pye, Shakespeare a greater poet than the feeblest of his forgotten rivals, Pope than Ambrose Philips, Byron than "*the hoarse Fitz-gerald*"? There is only one answer possible. These men in respect of poetry had been made giants by nature; those were condemned by nature to live and to die dwarfs.

And the same inequality that exhibits itself in the domain of poetry will be found in every other domain of human effort. What can be more unequal than the gifts of different singers? In every school and university we see multitudes of young men and boys whose opportunities of learn- ing are not only similar but identical, but of whom, in respect to assimilating what they are taught, not one in ten rises appreciably above a certain level, and not one in a hundred rises above it signally. We have Virgil at one end of the scale, and Bavius and Mævius at the other; at one end Patti, and the other the vocalist of the street; at one end a Scaliger and a Newton, and at the other the idler and the dunce, who can hardly conjugate τυπτω or stumble across the Asses' Bridge. And in practical life the same phenomenon repeats itself. Let us take any department of social activity or production, on the results of which the welfare of society at any given time depends. Let us take, for instance, the work of government, or invention, or commercial enterprise. In each of these we shall find a large number of men, each doing what is in him to subserve some particular end; and we shall find a few producing results which are great both for themselves and others, and the many producing results which are uniform in their individual pettiness.

It is perfectly true that in these great departments of practical life there may not be so obvious or so widely extended an equality of opportunity as that which prevails amongst poets, or amongst scholars in the same seminary, but in each department there will be a large number, at all events, whose opportunities are as equal as human ingenuity could make them. This is so in the

Enough men, as it is, have equal opportunities, to show how unequal men are in their powers of using them.

French army, in the English House of Commons, and in the world of business and industry ; and yet of men thus equally placed we see some doing great things, and doubling their opportunities by using them ; others doing little or nothing, and throwing their opportunities away. We have accordingly in every domain of activity a sufficient number of persons with the same external advantages, to show by the extraordinary difference between the results accomplished by them how great the natural inequality between men's capacities is, and how far the efficiency of a few exceeds that of the majority. It is therefore nothing to the purpose to attribute, as many reformers do, men's inequality in efficiency to the fact that equality of opportunity is not at present as general as it theoretically might be. To extend this equality further might produce good results or bad ; but in neither case would it tend to make men's capacities equal. The utmost it would do in this particular respect would be merely to widen the area of their realised inequality — to increase the number of the mountains, not to produce a plain.

No doubt a man may be ordinary in one respect, and great in another ;

It will doubtless be objected by those who would minimise natural inequalities that a man may be contemptible in one capacity—that of a poet, for instance—and yet be greater as a man than men who in one capacity are superior to him. It may, for example, be said that Frederick of Prussia, in spite of his

bad poetry, was a greater man than Voltaire. This
is perfectly true ; but it is necessary to explain clearly that it in no way contradicts what is being here asserted. It is, on the contrary, part of it. It cannot be too emphatically said that greatness, in the only sense in which we are here considering it—that is to say, as an agent of social progress—is a quality which we attribute to a man not with reference to his whole nature, but with reference solely to the objective results produced by him, so that in one domain of activity a man may be great, in another ordinary, in another decidedly stupid. What, then, we here mean by a great man is merely a man who is superior to the majority in his power but the majority are of producing some given class of result, whereas the not great in average man and the stupid are not superior to the any. majority in their powers of producing any.

The reader must thus entirely disabuse himself The measure of a man's of the idea that greatness, as an agent of social greatness as progress, has any necessary resemblance to great- an agent of social progress ness as conceived of by the moralist. A man may is the overt results actually be a great saint or a noble "moral character" who produced by passes his life in obscurity, stretched on a bed of him. sickness, and incapable even of rendering the humblest help to others. He is great in virtue not of what he does, but of what he is. But great-ness, as an agent of social progress, has nothing whatever to do with what a man is, except in so far as what he is enables him to do what he does. If two doctors were confronted by some terrible epidemic, and the one met it by tending the poor

Book II
Chapter 1

for nothing, and died in his unavailing efforts to save his patients, whilst the other fled from the infected district, and solacing himself at a distance with a mistress and an excellent cook, invented a medicine by which the disease could be warded off, and proceeded to make a large fortune by selling it, though the former as a man might be incalculably better than the latter, the latter as an agent of progress would be incalculably greater than the former.

A selfish doctor, if successful, is *greater* than a devoted doctor, if unsuccessful.

Again, if two doctors tried to invent such a medicine, and whilst the first succeeded the second failed, the second, though he might have exerted himself far more than the first, and have failed only owing to some minute flaw in his faculties, would be not only less great as an agent of progress than the first, but he would not be practically an agent of progress at all, any more than a man is an agent in saving another from drowning if he merely stretches a hand which the drowning man cannot reach, and actually himself tumbles into the water in doing so.

The fact that many men who produce no social results seem better and more brilliant than many men who do produce them, makes some argue that these results require no greatness for their production.

This truth, which sounds brutal when plainly stated, but is really little more than a sociological truism, is constantly overlooked, and even indignantly denied, by thinkers whose emotions are more powerful than their minds. The way in which such persons reason is very easily understood. They see that a number of men by whom great social results are produced—men who make successful inventions and who found great businesses—are narrow-minded, uncultivated, and contemptible in general conversation, and that a number of other men who produce no such results are scholars, critics, thinkers, keen judges of men and things ; and contrasting the brilliancy of those who have produced no great social results with the narrow ideas and dulness of those who have produced many, they proceed to argue that great social results cannot possibly require great men to produce them ; or, in other words, that they might be produced by almost anybody.

But the whole of this class of objections will altogether disappear when we more closely examine what the qualities are on which the production of given social results depends.   Let us take a few of these results as examples.   Let us take the formulation and the popularising of some particular political demand, by which the whole course of a country's history is affected, and the increasing and cheapening the supply of some articles of popular consumption—sugar, let us say, or workmen's boots and clothing.   The persons who urge the objections we are now discussing assume that all greatness, other than physical strength and dexterity, must be necessarily ethical or intellectual, and be calculated to excite our ethical or intellectual admiration.   But let them consider the qualities requisite to produce such results as have just been mentioned, and they will see that no assumption could be more wide of the truth.

A man who should, without underpaying his employees, succeed in manufacturing for the poorer

A lofty im-
agination is
often the
enemy to
practical
efficiency ;

classes boots, jackets, or shirts better in quality and very much less in price than those which they are accustomed to buy now, would probably have to devote a large part of his life to the consideration of a particular kind of seemingly sordid detail. To a man of wide culture and brilliant imagination, the concentration of his faculties on details such as these would be impossible ; and if he wished to produce any of the results in question, he would soon discover that he could not. The men who do produce them are rendered capable of doing so, not by the width of their minds, but by the exceptional narrowness. The intellectual stream flows strongly because it is confined in a narrow channel, and thus what to the superficial observer seems a sign of their inferiority, is really, so far as the results are concerned, one of the chief causes of their greatness.

> The mean man with the little thing to do
> Sees it and does it ;
> The great man with the great end to pursue
> Dies ere he knows it.

Robert Browning very tersely puts the case thus. We have only to alter his language in one respect. Seeing that the results we have now in view are realised results or nothing, the "*mean man,*" as an agent of material progress, will be the "great man," and the "great man" will be the little.

So, too, with regard to the man who affects

the history of his country by formulating and
popularising some particular political demand —
the secret of such a man's success, in four cases and great
efficiency is
out of five, will be found to lie in the greatness, not often independent of
of his intellect, but of his will—in an exceptionally exceptional
intellect.
sanguine temperament, in exceptional courage and
energy, and very likely in an exaggerated belief in
his own nostrums, which, instead of being a sign
of great intellectual acuteness, is incompatible
with it.

No doubt social progress, as a whole, has re- Intellect *is*
required for
quired and does require for its production intellectual progress, *e.g.*
in invention ;
powers of the highest and rarest kind. The point
here insisted on is that it is not produced by
intellectual powers alone, and that intellectual powers
alone would be quite unable to produce it. Thus
the sorrows and disappointments of the unfortunate
inventor are proverbial; and the reason is that great
inventive powers are frequently accompanied by a
very feeble will and a fantastic ignorance of the
world ; the inventor, though strong as a mind, being
pitiably weak as a man. He can do everything
with his inventions except make them useful to
anybody. He might be practically far greater were
he to lose some of his intellectual powers, could he but the inventor by him-
thereby develop some of the humbler qualities in self is often
helpless,
which he is wanting. As it is, he resembles a
chronometer which is without a main-spring, and
which is useless when compared with a ten-and-
sixpenny watch. Hence the inventor has so
frequently to ally himself with the man of enter-

Book II
Chapter 1

and has to
ally himself
with men
whose excep-
tional gifts are
unimpressive,
and even
vulgar.

prise, and only becomes great, as a social force, by doing so. Such unions are often sufficiently strange in appearance. We see some man whose intellect is the finest machine imaginable, but he is only redeemed from absolute and grotesque uselessness by his partner, who is little better than an inspired bagman. But such a bagman's gifts, however the inefficient theorist may despise them, are, though less striking than the inventor's, often quite as rare. No doubt many great inventors have the practical gifts as well as the intellectual, and their greatness, in such cases, is comprehended completely in themselves. It remains, however, an equally composite thing, no matter whether it takes two men or only one to complete it ; and exceptional intellect is only one of its elements. The other qualities with which it requires to be allied, and which alone give it its practical value, such as determination, shrewdness, and a certain thickness of skin, though often remarkable individually for the exceptional degree to which they are developed, just as often unite to produce practical greatness, not because of the exceptional degree to which they are developed, but of the exceptional proportions in which they are combined. Some of the most essential of them, indeed, need not be exceptional at all, except from the fact of their association with others that are so. Much greatness, for instance, of the most powerful kind, consists mainly of very ordinary sense in conjunction with extraordinary energy ; and energy is often, as has already been pointed out, in proportion

to the narrowness rather than to the width of the
imagination.

Greatness, in short, as an agent of social progress, Greatness is
is in most cases not a single quality, but a peculiar ^{not onequality,} but various
combination of many ; its composition varies combinations of many.
according to the character of the results in the
production of which the great men are severally
more efficient than the majority ; and it often
depends less on the extent to which any special
faculty is developed, in comparison with the same
faculty as possessed by ordinary men, than it does
on the degree to which each faculty is developed as
compared with the others possessed by the great
man himself.

When we speak of greatness, then, in the sense Greatness,
here attributed to the word—when we speak of then, is merely those qualities
great men as agents of social progress—we do not which, in any domain of pro-
mean that the world is divided into ordinary men gress, make the few more
and heroes. The members of that minority whom efficient than
we group together as great men, though some of the many.
them are, no doubt, of noble and heroic proportions,
are for the most part great in relation to special
results only ; even in relation to these special results
they are great in very various degrees, and many
of them in other relations may be ordinary,
or even less than ordinary. It must therefore be
clearly understood that greatness, as an agent of
social progress, is not an absolute thing, and that to
say of any one man that he possesses more greatness than another is a statement which, taken
by itself, has no definite meaning. When we

say that a man is great we mean that he is exceptionally efficient in producing some particular result, which is either implied or specified—that he is great in commanding armies, or in managing hotels, or in conducting public affairs, or in cheapening and improving the manufacture of this or that commodity ; and when we say that such and such a man possesses the quality of greatness to such and such a degree we mean that he produces results of a given kind, which are in such and such a degree better or more copious than results of the same kind which are produced by other people.

The inequality of men, then, in natural capacity being an obvious fact, and the nature and the degrees of their inequalities having been now generally explained, we may re-state, with a meaning more precise than was formerly possible, the fundamental proposition implied in the great-man theory, when that theory is raised from a rhetorical to a scientific formula. Progress of an appreciable kind, in any department of social activity and achievement, takes place only when, and in proportion as, some of the men who are working to produce such and such a result are more efficient in relation to that class of result than the majority ; or conversely, if a community contained no man with capacities superior to those possessed by the greater

number, progress in that community would be so slow as to be practically non-existent.

We must now go on to inquire what is the

precise way in which the men who are superior to the majority bring progress about; and we shall find that, however various they may be in other respects, they all promote progress in a way that is fundamentally similar.

all promote progress in the same way.

9

CHAPTER II

IT has already been explained that the *great man*, as here understood, does not in any way correspond with the *fittest man* in the Darwinian struggle for existence. The fittest man in the Darwinian sense merely promotes progress by the physiological process of reproducing his slight superiorities in his children, and thus raising in the slow course of ages the general level of capacity throughout subsequent generations of his race. The great man, on the contrary, promotes progress, not because he raises the capacity of the generations that come after him, but because he rises individually above the general level of his own. This, however, is only one of the differences by which the great man is distinguished from the fittest. There are two others, of which the first that we must consider is as follows.

In order to see how the great man promotes progress, we must consider that whilst the fittest survivor only promotes it

The fittest man, or the survivor in the Darwinian struggle for existence, is, so far as his own contemporaries are concerned, greater than his inferiors only in respect of what he accomplishes for himself, or for those immediately dependent on him. He is the man who lives and thrives whilst others die or languish, because whilst they can secure for themselves but little of what is requisite for life and health, he, by his superior gifts, is able to secure much. "*Families,*" says Mr. Spencer, "*whom the increasing difficulty of obtaining a living does not stimulate to improvement in production are on the high road to extinction, and must ultimately be supplanted by those whom the difficulty does so stimulate.*" That is to say, Mr. Spencer, and all our modern sociologists with him, conceive of the fittest as a man, or a man and his family, who fight for their food in isolation, like a lion and lioness with their cubs, and who affect their contemporaries only by being better fed than they, or as a race-horse affects its competitors only by being first at the winning post.

[marginal note: by living whilst others die,*]*

But the great man, as an agent of progress, shows his greatness in a way precisely opposite to that in which the fittest man shows his fitness. This it is that our contemporary sociologists all fail to perceive, and endless error is the consequence. The great man, unlike the strongest lion, promotes progress by increasing the food - supply not of himself, but of others ; or if he increase his own, as he no doubt generally does, he does so only by showing others how to increase theirs. He is like a lion who should be better fed than the rest of the lions in his region, not because he took a carcase from them for which they all were fighting,

[marginal note: the great man promotes progress by helping others to live.*]*

Book II
Chapter 2but because he showed them how to find others which they never would have found unaided, and took for himself in payment a small portion of each.

He promotes progress not by what he does himself, but by what he helps others to do.The great man, in fact, as an agent of social progress, is great not in virtue of any completed results which he produces directly, by the action of his own hands or brains, or which he exhibits in his own person, but in virtue of the completed results which, by some simultaneous influence which he exercises over the brains or hands of others, he enables others to exhibit in themselves, or produce or do in the form of products or social services.

In order to realise this great truth, let us begin with considering that form of greatness which promotes social progress by supplying it with its first materials, and from which all other kinds of greatness draw some portion of their nourishment.

We can see this by considering the progress of knowledge, which, as J. S. Mill says, is the foundation of all other progress.

It so happens that one of the most remarkable thinkers of this century, who, though he preceded Mr. Spencer, belongs to the same school, is able to assist us here by a very apt and remarkable passage. John Stuart Mill, in that section of his *System of Logic* to which he gives the title of "*The Logic of the Moral Sciences*," writes thus. "*In the difficult process of observation and comparison which is required (for the purpose of obtaining a better understanding of the laws of empirical sociology, and especially of social progress) it would evidently be a great assistance if it should happen to be the fact that one element in the complex existence of social man is pre-eminent over all the others, as the prime agent of the social movement. For we could then take the progress of that one element as the central chain, to each successive link of which, the corresponding links of all the other progressions being appended, the succession of facts would by this alone be presented in a kind of spontaneous order, far more approaching to the real order of their filiation than could be obtained by any other merely empirical process. Now the evidence of history and that of human nature combine, by a striking instance of consilience, to show that there really is one social element which is predominant and almost paramount amongst the agents of social progression. This is the state of the speculative faculties, including the nature of the beliefs which by any means they have arrived at, concerning themselves and the world by which they are surrounded. Thus,"* Mill continues, *"to take the most obvious case, the impelling force to most of the improvements effected in the arts of life is the desire for increased material comfort; but as we can only act on external objects in proportion to our know-ledge of them, the state of knowledge at any given time is the limit of the industrial improvement pos-sible at that time, and therefore the progress of in-dustry must follow and depend upon the progress of that knowledge."*

Any one who was inclined to be hypercritical might object, and object with justice, that the practical application of knowledge often lags be-hind the speculative attainment, and that material progress therefore, at certain times, depends on some new state of the practical rather than of the speculative faculties; but apart from this not very important inaccuracy of expression, Mill's way of putting the case is admirable for its lucidity and for its truth; and we may, for our present purpose, be content to take it as it stands. All civilisation depends on the accumulation of speculative knowledge, and all progress in civilisation depends on an increase in speculative knowledge.

But all pro-
gress in know-
ledge is the
work of
"decidedly
exceptional
individuals,"
Speculative knowledge, however, does not increase of itself. It is not acquired without considerable effort; and people acquire it only because they strongly desire to do so. Such being the case, let us turn to another passage, taken likewise from the writings of Mill, and occurring in the very same chapter as that which has just been quoted. "*It would be a great error*," says Mill, "*and one very little likely to be committed, to assert that speculation, intellectual activity, the pursuit of truth, is amongst the more powerful propensities of human nature, or holds a predominating place in the lives of any save decidedly exceptional individuals. But notwithstanding the relative weakness of this principle among other sociological agents, its influence is the main determining cause of social progress, all the other dispositions of our nature which contribute to that progress being dependent on it for accomplishing their share of the work.*"

Now what does this passage mean? About its meaning, and the truth of its meaning, there can be no possible doubt; but it will be well to observe

the extraordinary confusion in which Mill involves
what he means by his perverse manner of expressing it. In the first sentence of this last passage he tells us as clearly as possible that with regard to the pursuit of truth, and the power of discovering and understanding it, mankind are divided broadly into two classes—the great majority with whom the *"pursuit of truth"* and *"intellectual activity"* are *"slight propensities,"* and *"the decidedly exceptional individuals"* with whom these propensities are overmastering. But he has no sooner drawn this clear and all-important distinction between the two classes than he proceeds to undo his own work and mixes them together again in one unmeaning blur. He converts his statement that only *"the decidedly exceptional individuals"* desire truth with any great intensity, and have the faculties requisite for discovering it, into the statement that if we take *"the decidedly exceptional individuals"* and the majority together, and regard them as one body, which he calls *"mankind,"* we shall find that the average desire for truth is lukewarm, and the faculties for discovering it insufficient. He might just as well group Shakespeare with a hundred ordinary men ; tell us that Shakespeare could write the greatest poetry the world has ever known, and that the hundred other men could write no poetry at all, and then convert these statements into the following—that the one hundred and one men, Shakespeare included, could only write poetry of a very moderate quality.

This confusion of statement, however, on the part of Mill, is merely mentioned here in passing, as one more example of the nature of that inveterate error — namely the ignoring of the differences between one class of men and another—which has made modern sociology so useless for practical purposes. The sole point which really now concerns us is this. In spite of the verbal, and indeed the mental confusion into which Mill lapses, the truth which he was struggling to express, and which no one, he says, would be likely to contradict, is not that, as he nonsensically puts it, the speculative faculties are weak in mankind generally, but that amongst the larger part of mankind they have hardly any efficiency at all, whilst "*in decidedly exceptional individuals*" they are intense, active, and conquering; and that consequently it is these "*decidedly exceptional individuals*" who practically constitute "*the one social element which is predominant, and almost paramount, amongst the agents of social progression.*"

Now such being the case, let us resume our present inquiry, and ask how do these individuals who alone strongly desire truth, and have the faculties for discovering it, perform the practical part which Mill so rightly assigns to them? By what kind of conduct do they become "*agents of social progression,*" so as to raise communities from the level of helpless savagery and gradually endow them with all the resources of civilisation? One thing is perfectly clear. They do not so by the mere act of acquiring knowledge, by laying up this treasure in a napkin, or by showing it secretly to one another. They do so only by diffusing it, in such measure as is practicable, amongst a circle of men much wider than themselves. They do so, that is to say, by influencing the minds of others, by guiding their attention to this and to that fact, by providing, as it were, a go-cart for their weaker intellectual faculties, and compelling them to confront and assent to such and such propositions. All that mass of developing knowledge and expanding ideas which forms not only the basis but a part of all progressive civilisation, and is commonly called by the general name of enlightenment, is produced solely by the influence on average minds of the minds that are "*decidedly exceptional.*" It is not produced by the fact that the "*decidedly exceptional*" minds are stocked with such ideas and with such knowledge themselves, but by the fact that they communicate such a measure of these to average minds as average minds are severally able to receive.

To realise the truth of this we need do no more than consider for a moment the ordinary process of education. The schoolmaster and the college tutor, by the State or some other authority, are compelled to give their pupils instruction in certain subjects. But there is another kind of compulsion involved in the matter also ; and this has to do not with the selection of the subjects that are to be taught, but with what is to be taught about them. The general progress of a community depends primarily upon this; and what is to be taught about them is determined not by the State, or by any other legally constituted body, but by the masters of speculative knowledge, by contemporary men of science, scholars, historians, and philosophers. Knowledge advances because these men are not only adding to it, but because they are perpetually assimilating the new discoveries with the old; and these men, by means of their comments on previous writers, or by new works of their own, often reproduced in the form of text-books, put the word into the teachers' mouths; and the teachers, like the prophet Balaam, are compelled to speak it. In other words, great speculative thinkers are great as agents of mental civilisation and enlightenment only because, and only in so far as, they settle for others what these others shall believe and think.

A similar thing
is true of in-
vention, which
is knowledge
applied.
And now let us pass from mental progress to material—that is to say, from speculative knowledge to applied knowledge; and the truth that is being here insisted on will become clearer still. The master of knowledge, as applied to production, is the inventor. Now the most perfect and important machines ever devised by man—let us say the steam-engine and the printing press—had they been planned by their original inventors in all their present completeness, but kept by the inventors to themselves in the form of working models, made by their own hands and shut up in their own rooms, would have left the arts of life totally unaffected; our fastest means of travelling would still be the stage-coach;

our few books would be produced by the methods of the mediæval scriptorium. These machines are instruments of social progress only because, and in so far as, they are multiplied and brought into use ; and they could not be multiplied—as efficient implements, they could not be even made—without the co-operation of an enormous number of workers. It is probable indeed that in constructing the very model itself an inventor will have to employ some labour besides his own. Thus this first and preliminary step towards rendering his apparatus a factor in social progress he can take only by influencing one or two other men, at all events— artisans whose technical action he directs in such a way that it produces something specifically different from anything which it had produced before ; and as the apparatus is reproduced on a larger scale, put on the market, multiplied so as to meet a growing demand, and thus actually produces an effect on the arts of life, this practical result takes place only because, and in so far as, the number of artisans whose action is influenced by the inventor increases. The inventor, in other words, is an agent of "*social progression*" only because the particularised knowledge of which his invention consists is embodied either in models, or drawings, or written or spoken orders, and thus affects the technical action of whole classes of other men, just as Mr. Spencer affects, by means of his manuscript, the technical actions of the compositors who put his treatises into type.

Invention promotes progress only because the inventor influences the actions of the workmen who make and use his machines.

Material progress, however, depends not only on the inventor and his machine. It depends also on the uses to which his machine is to be put. Here we shall find a new kind of greatness to be necessary—that which is called business ability; and we shall find that this operates precisely like the greatness of the inventor, through the influence which its possessor exercises over other men.

The man of business ability promotes progress also only by so ordering others that the precise wants of the public are supplied.

All progress or development in commerce and in the arts of production is in proportion to the correspondence in every place and season of the goods brought into the market with the contemporary wants of the buyers. If it were not for this correspondence of the economic supply with the demand, progress in production would not be social progress at all; for just as a community does not become materially civilised by the mere act of wanting what it cannot get, so it does not become materially civilised by being presented with what it does not want—clothes, for example, which it could not possibly wear, and books in an unknown language, which it could not possibly read, or diminutive houses and furniture fit only for dolls. Now in any progressive community the wants of the buyers are in constant process not only of development but fluctuation, and are rarely quite the same in any two localities simultaneously. In order, therefore, that what is supplied may be in correspondence with what is wanted, it is necessary that in each industry the nature of the commodities produced be continually modified by men with a special sort of knowledge of the world ; and also, since want, in the sense of efficient demand, depends on the price at which these commodities can be supplied, it is necessary, just as it is in the case of the manufacture of machinery, that the army of men whose labour is involved in producing them shall be subject to men who, by their powers of industrial generalship, will be able to reduce the cost of reproduction to a minimum. Every business, in fact, and every industrial enterprise, succeeds or fails, not according to the amount of average labour involved in it, but according to the talents and energy by which this labour is directed. Thus in the economic domain, even more than in the intellectual, the great man is seen to be an agent of "*social progression*," in virtue not of the results which he himself produces by the direct action of his own hands or brain, but of the results which, being what he is, he causes to be produced by others.

And now having dealt with the great man as an agent of speculative progress which, as Mill says, is at the bottom of progress of all other kinds, and having dealt with him also as an agent of that manufacturing, commercial, economic or material progress which Mill cites as the chief example of what practical progress is, and having shown how the essence of his greatness is his power of influencing others, let us illustrate this truth finally by a brief reference to three other kinds of human and social activity which exhibit it

And the same principle is obviously true in the domains of war, politics, and religion.

in a light so obvious that it requires no explanation. These three kinds of activity are the military, the political, and the religious. The great soldier, as has been said already, is essentially the great commander—the man who makes others act and group themselves in a specific way. The statesman not only aims at benefiting his countrymen generally, but he achieves his aim by the same means as the soldier, namely, by influencing the actions of others in certain specific respects; whilst the man who is socially great in the domain of morals and religion is the man whose teaching and example affect the actions, and even the inmost feelings, of multitudes, or gives precision to their faith.

But here, having reduced to a truism this important truth that the great man, as an agent of social progress, is great only because he is able to exercise Greatness,
however, is not
in all cases
equally bene-
ficial. a specific influence over others, it is necessary to turn our attention to a different order of facts altogether. Greatness, as we have seen already, is of very many kinds. It is a varying compound of various and variously developed qualities; and its degree is measured by its efficiency in producing this or that result by which society is benefited. But greatness, in the sense of exceptional power of so influencing others that some given result shall be produced by them, has other varieties besides those that have been already mentioned. Each domain of progress has not only its own leaders, but it has leaders who desire to lead men in very different directions. There are scientists

with conflicting theories, inventors with rival inven- tions, statesmen with rival policies. It follows accordingly that though all these men may be pos- sessed of talents indefinitely above the average, they would not all of them, were their influence over other men equal, affect society in an equally advantageous way. Some men, indeed, whose talents are "*decidedly exceptional*" would, on account of some flaw or defect in their character, not promote, but, on the contrary, retard true progress, in exact proportion as they made their views prevail. Thus, though all progress is due to great men, all great men would not promote progress; or they would, at all events, not promote it equally. Progress, therefore, as resulting from the actions of great men, depends on the degree to which certain of them make their own views prevail, and secure the rejection of others which are directly or indirectly opposed to them. It depends, that is to say, on a keen competitive struggle which is continually taking place within the limits of the exceptional minority.

And here we come to that further point of difference, which still remains to be noticed, between the part played in social progress by the great man, and the part in it played by the fittest according to the Darwinian theory. Two points of difference between them have been noted and explained already, one being that the fittest man promotes progress only because he raises, by a physiological process, the average capacities of his successors, whereas the great man promotes pro-

[right margin notes:]

Book II
Chapter 2

The influence of some great men is more advantageous than that of others.

Progress, then, involves a struggle through which the fittest great men shall secure influence over others, and destroy the influence of the less fit.

We now come to another point of difference between the fittest great man and the fittest survivor.

gress because he is himself more capable than his contemporaries ; the other being that the fittest fulfils his social function by fighting for his own hand, without any reference to others, whereas the great man fulfils his solely by influencing others. We are now coming to a third point, which is, for practical purposes, even more important than the preceding.

The great-man theory, just like the theory of Darwin, involves a competitive struggle. This struggle is a struggle between great men ; and its existence is a fact of too obvious a character to have escaped the notice of even the most inaccurate of our social evolutionists. But they one and all of them have completely misunderstood its nature. They have hastened to identify it with the Darwinian struggle for existence, from which it differs in the most vital manner conceivable ; and, obscuring it thus by a loose and misleading analogy, they have managed to blind themselves to its entire practical significance. The Darwinian struggle for existence no doubt has its counterpart in the contemporary competition of labourers to find remunerative employment, and in the fact that those who are least successful in finding it would, if left to themselves, be continually dying off. In a progressive country there is, or there always tends to be, a larger number of would-be labourers than there is of tasks which at the moment can be profitably assigned to them. A struggle therefore is involved in obtaining work of any kind ; and for the higher kinds of work the struggle is very keen. But this is not the

The social
counterpart to
the Darwinian
struggle for
survival is to
be found in
the struggle of
labourers to
find employ-
ment.

struggle to which modern progress is due. Progress, in the sense of the rapid and appreciable
movement which alone concerns us here, is—to
confine ourselves for a moment to the domain of
industry—not the result of a struggle to execute
work in the best way, but is the result of a struggle
to give the best orders for its execution. It presupposes the existence of a certain amount of skill ;
but it does not, except in its very earliest stages,
depend on the struggle of so many thousand men,
each to become individually a more skilful worker
than his fellows. It is, on the contrary, when its
earliest stages have been passed, so independent of But this is not
the struggle
any further increase of skill in the individual worker, to which historical progress
that it continues its course whilst skill remains is due;
stationary.

This is shown by the fact that some of the greatest
advances ever made in material civilisation have
been made during the active lifetime, and with the
aid of the hands and muscles, of a single generation for the most
rapid progress
of workers, and has implied no improvement at all has taken place
without any
either in their acquired faculties or their inherited. increased fit-
Let us take, for instance, the introduction of the ness in the
labourers.
electric light, and the way in which it is superseding
gas. The mechanics first employed to make the
appliances for its production were none of them
asked to perform any task which required on their
part any new knowledge or dexterity. All they were
asked to do, and all they did, was to submit their
existing faculties to some new external guidance :
and the electric light, in so far as it has superseded

10

gas, has superseded it not because it is the product of more skilful labour, but because it is the product of manual labour directed by a set of inventors and employers, who, so far as regards certain social requirements, direct it more successfully than another set. The struggle which it represents is a struggle between employers only. It does not, except by accident, represent any struggle between the employed.

And what is true of the struggle which produces industrial progress, is true of that which produces progress of all other kinds. Scientific knowledge increases in proportion as those exceptional individuals whose studies have brought them most near to the truth are able to fight down the opinions of the exceptional individuals who differ from them, and to impress their own undisputed upon the world. Such knowledge does not increase on account of any struggle amongst the learners, which causes some of them to become more and more apt in learning. It grows on account of a struggle between philosophers, each of whom aims at settling what the learners shall learn. And with regard to religion and politics the case is just the same. The progressive struggle is primarily between rival prophets and politicians. The spread of Christianity, for instance, was not brought about by Christian races exterminating those that were not Christians. It was brought about by Christian thinkers and teachers discrediting the doctrines taught by thinkers and teachers who were opposed to them. Free-trade, again, in this country has not triumphed over pro- tectionism, because the mass of free-traders have exterminated the mass of protectionists.   It has triumphed simply because, in the eyes of the majority, one school of theorists has succeeded in discrediting another.

Now these facts, which, when once stated, are so obvious, not only throw the Darwinian struggle for existence altogether into the background as an agent in social progress, but they show that it presents us with no true analogy to that kind of struggle from which progress principally results.   They show us, on the contrary, that the struggle which produces social progress, though it resembles the Darwinian struggle in one point, is in all other points contrasted with it.   The struggle of one employer against another to direct labour in the most advantageous way, or the struggle of one politician or religious teacher against another to secure for his own views the largest number of adherents, is so far like the Darwinian struggle for existence, that it is a struggle in which individual is pitted against individual, and the gain of the successful is the loss of the unsuccessful.   But the limits within which this struggle is confined are very narrow indeed ; and the mass of the community takes no part in it whatsoever.

In the progressive struggle between great men, the mass of the community play no part whatever.

In order to show this with the utmost clearness possible, let us turn again to the domain of economic progress, which generally supplies the sociologist with his simplest and most luminous illustrations.

Let us take,
for instance,
two rival hotel-
keepers. The success of the strongest and ablest employers— that is to say, the heads of the most successful businesses—may involve, and does involve their selection for survival as employers, and does involve the extinction, as employers, though not necessarily as men and parents, of their weaker and less able rivals ; but it involves no struggle for existence with the men employed by them—that is to say, with the great masses of the community. Two men, we will say, start rival hotels, and each begins with a staff of a hundred persons. One of the two understands One becomes
bankrupt, and
the other takes
over his hotel
and his staff. his business far better than the other. His hotel is always full, whilst his rival's is half empty. The latter at last becomes bankrupt ; the former buys his business, and together with his premises takes over his staff. He employs two hundred persons, instead of a hundred as at first ; the hotel of the bankrupt, which the bankrupt ran at a loss, now yields the same profit as the other ; and the aggregate takings of the two are thus increased largely. Here we have a community of two hundred and two persons offering a marked example of great material progress ; and this progress has been the result of a genuine struggle for existence. But the struggle for exist- ence has been between two persons only—that is to say, between the two hotel-keepers. As *hotel-*
keepers existence is the very thing they have been struggling for, and the survival of the one has meant the disappearance of the other ; but between them and the two hundred persons employed by them there has been no struggle at all. The achievement

by the successful hotel-keeper of a fortune double
that with which he started has not involved any
diminution in the wages of his staff. It will, on the
contrary, if we are to take the case now in question
as typical of the survival of the fittest employers
generally, have not only not diminished their
wages, but very largely increased them. For here
there is one further truth which naturally introduces
itself to our observation. Whatever allowance it
may be necessary to make for the lowest class or
residuum of our modern populations, it is the most
clearly proved and prominent fact in modern industrial history—and one which even socialists are now
ceasing to deny—that along with the vast increase
in wealth which the ablest employers have, by their
struggle with rivals, secured for their own enjoyment, there has been not a corresponding diminution, but a corresponding increase in the means
of subsistence that have gone to the population
generally. The average income per head in this
country of that class—composed mainly of wage-
earners—which does not pay income tax has, in
terms of money, nearly trebled itself during the
present century ; its purchasing power has increased .
in a yet larger ratio, and its increase will be found to
have been most rapid and striking at periods when
the struggle amongst the employing class has been
keenest.

It will thus be seen that the struggle which produces economic progress—and progress of every
kind is produced in the same way—is not a general

Book II
Chapter 2

The staff of
the unsuccess-
ful hotel-keeper
gain, not lose,
by being em-
ployed by the
successful.

Historical pro-
gress, then,
results from a
struggle

struggle which pervades the community as a whole ;
neither is it a struggle between the majority and an
exceptionally able minority, in which both classes
are struggling for what only one can win, and in
which the gain of the one involves the loss of the
other ; but it is a struggle which is confined to the
members of the minority alone, and in which the
majority play no part as antagonists whatsoever. It
is not a struggle amongst the community generally
to live, but a struggle amongst a small section of
the community to lead, to direct, to employ, the
majority in the best way ; and this struggle is
an agent of progress because it tends to result, not
in the survival of the fittest man, but in the
domination of the greatest man.

CHAPTER III

THE whole secret of social progress, other than the most rudimentary, is summed up in the formula with which the preceding chapter has concluded. Progress is the result of the domination or the triumphant influence of the greatest. That is to say, the civilisation of the entire community depends alike for its advance and for its maintenance on a struggle which is confined within the limits of an exceptional class ; and the ordinary members of the community are connected with it only by the fact that when the fittest competitor achieves the domination for which he is struggling, they, instead of being defeated by him, share the advantage of his victory. When the scientific doctor discredits the theories of the quack, when the competent organiser of industry causes the ruin of the incompetent, when a good ministry drives a bad from office, when a great general supersedes one who is inferior, or when a true religious teacher destroys the influence of a false, the whole community gains, except the men who have personally lost

All gain by the domination of the fittest, except the few who fail to secure power for themselves.

authority, and who share the merited fate of their own errors or deficiencies.

The progress and the maintenance, then, of civilisation in any community depends on its possessing a number of great men, of which number the greatest shall, by competition with the others, succeed in gaining a control over the beliefs and actions of the majority.

We must consider, however, that the great men who struggle for domination would not do so without some strong motive ; Here, however, we are introduced to two new sets of facts, which have not thus far come under our consideration at all.

In the first place, great men do not come into the world ready-made. Their greatness is potential only, or in other words it is practically non-existent, until it has been developed ; and the process of developing it is in most cases extremely arduous. The philosopher, the soldier, the inventor, the statesman, the great merchant or manufacturer, achieve success only by prolonged and intense effort, by study, by concentrated thought, by action, by rude experience. Genius, indeed, has been defined as an infinite capacity for taking trouble ; and the definition, though very incomplete, is, so far as it goes, true. No one, however, takes trouble without a motive ; and a motive being some object of desire, such as money, rank, or pleasure, which a man hopes to attain by a certain line of action, it follows that if a community is to possess great men as actual agents of progress, and not merely as wasted potentialities, its social constitution must be such as to offer and make attainable positions, possessions, pleasures, or other advantages which its potentially great men will feel to be worth working for.

In the second place, since the great man, as we have seen, is an agent of progress and civilisation only because he influences others—because he guides their speculative beliefs, and in certain respects commands their actions—the society or community to which the great man belongs must be such as not only to supply him with a motive for exercising this influence, but also to enable him to secure for himself the means by which it may be exercised ; and, furthermore, the means in question must be of a kind which will enable the rival great men to bring their respective capacities to a decisive practical test, so that the influence of the most efficient may establish itself, and that of the less efficient cease.

and also that they cannot dominate others except by some particular means.

Now the whole question of motive we will deal with later on. We will for the present put it altogether aside. We will assume a natural impulse on the part of all great men to develop their powers to the utmost, and employ them in influencing others, wholly independent of any other reward than such a minimum of sustenance and comfort as is physically essential to their efficiency ; and we will confine our attention altogether to the question of the means by which the influence of the great men over the majority is obtained.

Now the question of motive we will treat of hereafter. At present we will confine ourselves to the question of means.

Human progress, however, being a complex thing, and taking place in different domains of activity, the means by which the great man influences others will vary with the nature of the results which his

These vary in each domain of social activity.

influence aims at eliciting. The social activities on which progress depends, though they may be sub-divided indefinitely, are reducible to five kinds—intellectual, religious, military, economic, and politi-cal ; and with regard to the two first, the influence of the great man exerts itself to determine what others shall believe and think ; with regard to the three last, it exerts itself to determine what others shall do.

Now out of these five domains of activity the three first—namely, the intellectual, the religious, and the military — are such that the means by which the great man makes his influence felt in them hardly require discussion. In the first place, they are obvious—there is no dispute about what they are ; and, in the second place, the fact of their being what they are has no bearing, except such as is very remote, on any disputed question concerning the practical organisation of society. In the in-tellectual world thinkers, scholars, and men of science gain their influence by discussions, for the most part embodied in books, which discussions are carried on before a jury of expert critics, each man defend-ing his own views against the views of those who differ from him ; and the jury of experts ultimately gives its verdict, to which sooner or later the com-munity at large submits. The religious leader gains his influence similarly. He gains it by arguments and persuasions, which are felt by a band of followers to touch the spirit more deeply than those of other prophets. He gives to his disciples, and his

disciples give to the multitude. But these means
are of so universal a kind, and have so little connection with any specific social arrangements, that
none of the disputed points of social politics are
involved in them ; and we consequently have at
present no occasion to discuss them. So, too, with
regard to the military leader, though the means which
are employed by him do, beyond a doubt, imply
social arrangements of a very specific kind—namely,
an iron system of discipline, with death and the lash
to sanction it ; yet these arrangements, however they
may be denounced by sentimentalists, have always
been found essential to the efficiency of every army ;
and though many worthy people would abolish
military activity altogether, and whilst socialists
especially express themselves anxious to do so, it
is perfectly evident—nor would any socialist deny
it—that a socialist State, if it had to fight for its
existence, would be obliged to enforce the required
military discipline by methods essentially identical
with those of Cæsar or Wellington. It may, indeed,
be disputed whether the great military leader is
not a superfluous figure on the social stage ; but
so long as his greatness makes itself felt at all,
it will continue to make itself felt by the same
means.

The only domains of social activity, therefore, in We need con
sider what th
are only in
the domains
of politics an
of wealth-pro
duction.
which the means employed by the great man to
control the actions of others so that ordinary men
may be guided by the faculties of the exceptional—
the only domains of activity in which these means,

thus employed, really require minute and careful discussion, and have really a direct bearing on the practical problems of the day—are the domain of economic production and the domain of political government. These, indeed, may be said to contain between them the whole of the questions with regard to which parties are divided—with regard to which those who believe that the conditions of civilisation may be indefinitely improved but can never be fundamentally altered, are divided from those who believe them to be capable of indefinite metamorphosis.

1e question
most im-
rtant in its
arings on
alth-pro-
ction.
This is specially true of the domain of economic production ; for it is mainly on account of its connection with the production and distribution of wealth that political government excites so much popular interest and forms the subject of so much vehement controversy. And in every other domain of human activity equally, we shall find that the interests, the endeavours, and the disputes of men have an economic process as their basis, or economic progress as their object. The processes of production and commerce are, in fact, the central processes of every nation's life. Government exists to foster them, and changes its form as these processes develop; whilst fleets and armies exist mainly for their protection, and more and more depend on the progress that takes place in them. It is, in short, in the domain of economics that all the social problems of the day either begin or end ; and consequently in examining the means by which the great man influences others, the question which it is

really our first concern to examine relates to the means
by which great men, whose greatness consists in the fact that they are exceptional in their powers of causing the production of wealth, and on whom consequently the wealth of the whole community depends, obtain a control over other men's productive actions.

This control can be secured in two ways only, or else in some way that is a combination or modification of both. One of these ways is slavery; the other is the capitalistic wage-system. Let us consider how the two resemble each other, and also how they differ.

The great man in wealth-production can influence the action of others by two means only—by the slave-system and the wage-system.

They resemble each other because both, in so far as they subserve progress, subserve it for precisely the same reason. They are both contrivances by which the superior few may secure, so far as industry is concerned, the implicit obedience of the many. On the private lives of the many their effects will be widely different; but so far as concerns their direct connection with industry—their operation on men during the actual processes of production—slavery and the capitalistic wage-system differ only in this: that the one secures the required industrial obedience by operating on men's fears; the other secures it by operating on their desires and wills. Thus the slaves who built the pyramids had each some specified task—the making of so many bricks, the cutting of such and such stones, or the fixing of bricks and stones in such and such situations—which had to be performed if the pyramids were to be built at all. So, too, if the Hotel Metropole at Brighton was to

The slave-system secures obedience by coercion, the wage-system by inducement.

be built at all, the bricklayers, masons, and other workmen who built it had to perform tasks of a precisely similar kind. But obedience to orders on the part of the Egyptian slave was secured by the knowledge on his part that disobedience would be punished by some form of chastisement, and very likely of torture, whilst obedience on the part of the Brighton workman was secured by the knowledge on his part that, unless he chose to yield it, one way, at all events, of earning a livelihood would be closed to him.

It is this latter method of securing industrial obedience that is made possible by the capitalistic wage-system ; and it is primarily for this reason that what is called capitalism is an agent of progress, and has developed itself in progressive communities. As for capital itself, this, as we all know, performs part of its functions by assuming the form of machinery, buildings, bridges, railways, and a variety of structures and appliances which are grouped together under the general head of *fixed capital* by economists. But these structures and appliances are themselves the result of the previous influence of great men on the industrial actions of the many ; and as it was by means of wage-capital that this influence was secured, the primary and most essential functions which capital fulfils, and which really form the essence of the capitalistic system, are to be found by considering capital as employed in the payment of wages.

Now capital as thus employed consists of an

accumulation of the necessaries and comforts of life, by the consumption and use of which men are able to sustain themselves when engaged on works requir- ing a long period for their completion, which will when completed be useful and produce much, but which, until they are completed, will be of no use at all, and will consequently supply nothing to the workers when actually engaged on them. The simplest example of work of this kind is agriculture. The first man who saved sufficient food to support himself, whilst tilling the soil and waiting for his crops to ripen, was the first capitalist. But capital, when it takes the form of accumulated necessaries and comforts, though it now reaches the workers in the form of wages usually, need not do so of neces- sity. It need not do so when the work is extremely simple and the methods employed are rude. Where- ever agriculture, for example, is in its earliest stages, every husbandman may be his own capitalist, and start with an accumulation of food in his own cottage which will keep him alive till his crops are ready for sale or for consumption. In cases such as these we have capital which, so far as its substance is con- cerned, is identical with wage-capital, but is not wage-capital nevertheless. In order to turn it into wage-capital it is necessary that these accumulations of food shall pass out of the control of the workers —such as the husbandmen just referred to—and be brought under the control of some other person or persons, who will dole them out to the workers on certain conditions only. The wage-system, in short,

Wage-capital is an accumu- lation of neces- saries of life,

owned or con- trolled by a few persons,

does not represent capital as such.   It represents capital, in the form of the immediate means of subsistence, as owned or controlled by a small number of persons ; and its efficiency as a productive agent resides in the bargain which it enables any great and apportioned by them
amongst
many, on
certain con-
ditions. man possessing it to make with ordinary workers— a bargain, not that they shall work such and such a number of hours (for that they would have to do were each man his own employer), but that they shall do their work in accordance with the great man's directions.

Now this fact that the wage-system represents the control of capital by the few—and this is its essential characteristic—is the fact on which, more than on any other, the socialistic opponents of the modern wage-system insist.   They are never weary of insisting that it has its foundation in a monopoly. But though they perceive the fact, they entirely
miss its significance.   Karl Marx conceives of the capitalists as a body of men who, so far as production is concerned, are absolutely inert and passive. Owing to a variety of causes, he says, during the past four hundred years all the means of production have come under their control, and access can be had to them only, as it were, through gates, of which these tyrants hold the key.   Outside are the manual labourers, who are the sole producers of wealth, but who, without the means of production, naturally can produce nothing — not even enough to live on ; and the sole economic function which the capitalist fulfils is to let the

labourers in every day through the gates, on the condition that every evening the unhappy men render up to him the whole produce of their labours, except that insignificant fraction of it which is just necessary to fit them for the labours of the day following. Now it is no doubt theoretically possible that a society might exist, composed of a mass of undifferentiated and undirected manual labourers on the one hand, and on the other of a few passive monopolists who extracted from them most of what they produced, as the price of allowing them the opportunity of producing anything; but it is perfectly certain that a society of this kind would exhibit none of the increasing productive power which, as even Marx and his school admit, is one of the most distinctive features of industry under the capitalistic wage-system. Under that system productive power has increased, not because capital has enabled a few men to remain idle, but because it has enabled a few men to apply, with the most constant and intense effort, their intellectual faculties to industry in its minutest details. It has increased not because the monopoly of capital has enabled the few to say to the many, "We will allow you to work at nothing, unless you give us most of what you produce," but because it has enabled them to say to the many, "We will allow you to work at nothing, unless you will consent to work in the ways that we indicate to you."

The few, so far as our present argument is

The essence of these conditions is that the many shall be technically directed by the few.

Book II
Chapter 3

The question
of how much
the few
appropriate of
the product is
a separate
question
altogether.

concerned, may appropriate much of the gross product or little; or they may leave the whole of it to be divided amongst their employees. What they actually have done, or do, or may do, in this respect, is another question altogether, and will be discussed hereafter separately. The essence of the wage-system, in so far as it has influenced the actual processes of production, is in the power it gives to the few to direct the producers, not in the power it gives them to appropriate the products. It will indeed require very little reflection to show us that if the great men in the industrial world would only develop and use their faculties, without any motive of ambition or self-interest to stimulate them, — as indeed at the present moment we are assuming that they do,—they could use the wage-system for the purpose of directing industry merely by monopolising the control of capital without monopolising, and even without sharing in, its possession.

The *corvée*
system or
slavery would
make wage-
capital super-
fluous; and
then thus show
us what the
essential
function of
such capital is.

This truth will become plainer still when we reflect that if only certain conditions prevailed which in many civilised countries survived till quite recently, the whole process of production as we now have it might be carried on without any wage-capital at all. These conditions are those of the *corvée* system, under which peasants and others who owned the lands upon which they lived, and maintained themselves on those lands in a certain position of independence, were compelled to place their labour, for so many days a week, at the absolute

disposal of this or that superior.   Such a system, if
applied to modern industry, would have, no doubt,
many incidental disadvantages ; but if only a number
of independent peasant-proprietors could be forced
to give half their time to the proprietor of a
neighbouring factory, and during that time to work
in it under his orders, the entire use and necessity of
wage-capital would in theory, at all events, be gone.
The same thing is also true of slavery, between
which and the wage-system the *corvée* system stands
midway.   Like the peasant - proprietor, who is
forced to give part of his labour to his over-lord,
the slave is supplied with the necessaries of life
independently of his obedience to the· detailed
orders of his task-master.   The peasant maintains
himself by tilling his own fields ; the slave-owner
feeds his slave just as he would feed an animal.   In
neither case is the giving or the withholding of a
livelihood used as the motive or sanction by which
industrial obedience is ensured.   Obedience is
ensured by the direct application of force, or the
knowledge on the slave's part or the peasant's that
force will be applied if necessary.

It will, no doubt, be urged by some that whatever
assistance is afforded by the talents of the few to the
industrial efforts of the many, may be secured by
a third means, which is neither slavery nor yet the
wage-system—that is to say, by what is called the
system of "co-operation."   Co-operative production,
however, when it differs in anything except in name
from production as carried on under the ordinary

wage-system, differs from it only in being the wage-system under a thin disguise. For the ideal co-operative factory is simply a factory in which all the shareholders are workers, and all the workers are shareholders, and in which, being shareholders, they elect their manager. Under such conditions, each of these working shareholders may receive his remuneration under the form, not of wages, but of profits. But if any shareholder, or any group of shareholders, should systematically shirk working, or disobey the manager's orders, the whole, or a part of the payment that would be otherwise due to him, would be withheld; for unless some regulation of this kind were in force, it would be impossible to ensure any co-operation amongst the co-operators, or any order, or any equality of diligence. Each worker's profits, then, are in reality his wages, being essentially a payment which is made to him only on condition that he performs certain specified tasks in a certain specified way.

There are, then, only two alternatives— the wage-system and the slave-system; We are thus brought back to the point from which we started—namely, that there are two methods only by which, in the domain of industry, the superior faculties of the few can direct the faculties of the many : firstly, the capitalistic wage-system, which is the method of inducement ; secondly, slavery, complete or partial, which is the method of coercion. And of the truth of this assertion the reader shall now be presented with a highly interesting and curiously conclusive proof, taken from the very last quarter in which he would naturally expect to find it. This proof is afforded us by the schemes which, with ever-increasing clearness, have of recent years been put forward by all the more thoughtful socialists.

These enthusiasts, who are still careful to tell us that they regard the wage-system as the source of all social evils, have been slowly coming to perceive that the ability with which the labour is directed is as important a factor in production as the labour itself, which is directed by it. They propose accordingly to regenerate the human race by transferring the ownership of capital from private employers, not to groups of factory-hands, as the "co-operators" propose, but to the State; and by substituting for the private employers a hierarchy of State officials. Now these officials, so far as the wage-system is concerned, if they differed at all from private employers of to-day, would and could differ from them in the following way only. The present dispensers of wages assign the means of sub- sistence to each worker in proportion to the exactness, intelligence, and efficiency with which he obeys orders. The dispensers of wages under socialism would dispense these means daily to every worker alike, with no immediate reference to his in- dustrial actions whatsoever;. and the direction of his actions would be a second, and wholly distinct process.

That such is the case is shown, and indeed distinctly admitted, in a preface to the American edition of *Fabian Essays*. It is there stated that with regard to the apportionment of the means of subsistence, the only "*truly socialistic*" scheme is one which would "*absolutely abolish*" all economic distinctions, "*and the possibility of their again arising, by making an equal provision for the maintenance of all an incident and an indefeasible condition of citizenship, without any regard whatever to the relative specific services of different citizens.*"

For they would secure industrial obedience by coercion, *The rendering of such services, on the other hand, instead of being left to the option of the citizen, with the alternative of starvation, would be required under one uniform law or civic duty, precisely like other forms of taxation or military service.*"

Such, then, is the most advanced socialistic programme—the programme of the men who have set themselves to devise an escape from capitalism. An escape from capitalism it may be ; but it is an escape into complete slavery. For the very essence of the position of the slave, as contrasted with not through the worker's own desire to earn his living. And this is the essence of slavery. the wage-labourer, so far as the direction of his productive actions is concerned, is that he has not to work as he is bidden in order to gain his livelihood, but that, his livelihood being assured to him, he has to work as he is bidden in order that he may avoid the lash, or some other form of punishment ; and amongst all the more thoughtful socialists there is now a consensus of admission that the socialistic State would necessarily have in reserve the severest pains and penalties for the idle and the careless and the disobedient.

Since, then—let us once more repeat it—the

progress and maintenance of economic civilisation depend, as even socialists are now beginning to perceive, on the industrial actions of average men being subjected to the control of exceptional men, and since this control can be secured by two methods only—that of the wage-payer and that of the slave-owner—it is evident that all progress and civilisation implies the existence of either one system or the other, and that socialists accordingly, in proportion as they reject the wage-system, are obliged to replace it by what is essentially the system of slavery.

Book II
Chapter 3

We have thus far, however, dealt with but one half of our subject. We have considered merely the means by which any one great man exercises industrial control over the actions of a number of ordinary men. We have still to consider the means by which the most efficient of the great men get this control into their own hands, and take it out of the hands of the less efficient.

Next let us consider the means by which the great directors of industry compete against one another.

Under the *régime* of private capitalism this process is simple. The fitness or efficiency of each great man is according to the acceptability to the public of the goods or services which he offers them. If the public are not pleased with these goods and services, they do not buy or demand them ; and the capital of the man by whom they are offered, not being renewed by any money received, melts in his hands, and with it his control over other men's labour. Meanwhile, by a converse process, the great men who offer goods and services which the public desire

Under capitalism they do so, owing to the fact that the man who cannot direct industry so as to please the public, loses his capital, and with it the means of direction.

Book II
Chapter 3

and find serviceable, renew and increase by their payments the capital which has been disbursed by him, and renew and increase his control over other men's labour along with it.

The wage-system is the only efficient means of competition of this kind.

Now if the wage-system is the sole alternative to slavery as a means by which the great man controls the actions of the ordinary man, it is still more obviously the sole alternative to slavery as a means by which one great man, in controlling them, shall compete against another great man. Indeed, we may speak still more strongly. We may say not only that it is the sole alternative means, but that it is the sole efficient means. And if we desire a proof of this, all we have to do is to repeat our former procedure, and consider how the socialists propose to supply its place.

The socialists, though they affect to be opposed to competition altogether,

It is, no doubt, true that when we first begin this consideration it does not appear that we should derive from it much direct enlightenment; because, if we may go by what the socialists themselves tell us, one of their principal objects is to abolish competition altogether. Their protestations, however, with regard to this matter betray a most curious and most amusing confusion of thought. They declare that competition must be abolished because it inflicts misery on the majority—that is to say, on the weakest in what they call the "*cut-throat struggle.*" But, as was shown at great length in the last chapter, competition means two, and two absolutely distinct things — one being a struggle to live, the other a struggle to dominate; and

the effects of the two on the majority are
altogether different. To this fundamental truth,
the socialists are completely blind. The struggle
to live, or, in other words, the struggle to
secure employment, no doubt, when it is severe, does
entail suffering on the strugglers. But this struggle,
though it often accompanies progress, under the
capitalistic system is not essential to it — as is
shown by the fact that when such progress is
most rapid the struggle in question tends to disappear altogether ; for the competition is then
amongst the employers to find labour, rather than
amongst the labourers to find employment. Now if
the struggle for employment could be obviated by
any kind of social reform, an indubitable benefit
would, no doubt, be conferred on the workers
generally. But just as this struggle for work or for
existence — this struggle of one worker against
another—is not essential to the capitalistic wage-
system, and certainly did not originate with it, and
just as that system would not necessarily be
abolished by its overthrow, so it is not the kind of
competition against which the socialists direct their
main attacks. Their main attacks are directed
against the struggle between the wage-payers, not
the wage-earners—that is to say, against the struggle
not for existence, but for domination ; and the
struggle for domination has on the workers generally
no evil effects at all, except such as are occasional
and accidental. On the contrary, the workers are as
much interested in its maintenance as anybody ; for

not only does it inflict no injury on themselves, but to it that progress in the processes of production is due on which their own hopes depend, as much as do those of their employers. Accordingly, the socialists, profound thinkers as they are, propose to abolish the competition by which the workers benefit, because they confuse it with the competition by which the workers suffer. The point, however, which concerns us here is not that they have made a blunder as to the kind of competition which they

should attack, but that the kind of competition which they declare themselves pledged to abolish, as a thing accursed, and the root of all social evils, they really reintroduce into their own programme, altered only by being associated with the system of slavery, and by being robbed of its practical efficiency, and robbed of nothing else.

the only
change being
that it is
associated with
the slave-
system, which
is very
cumbrous and
inefficient.
For our contemporary socialists, who have at last come to perceive that the productivity of labour depends on the ability with which it is directed, perceive also the fact that, out of many possible directors, some would direct it far more efficiently than others. They also perceive the fact that the directors of labour, who, according to their proposals, would be officials of the bureaucratic State, could prove their efficiency only by practical experiment. Now if all capital were, as socialists propose it should be, owned by the State, and if all the means of subsistence were apportioned amongst the citizens equally, without reference to the work performed by them; and if all the directors of labour, whether inventors or business

organisers, had to act as State officials, or else not act at all, the practical experiments necessary to show which officials were the fittest could be brought about only by the State investing such and such of them with a quasi-military power over so many regiments of labourers for such and such a time, which power would be renewed if they could persuade the State to reappoint them, or taken from them if the State should be persuaded that some other men, their rivals, would employ this power more usefully. And this is precisely what the proposals of the socialists come to. The whole multitude of State officials who would direct socialistic industry would, according to every socialistic programme, be appointed, promoted, or degraded to the ranks of ordinary workers in accordance with the efficiency shown by them in the practical command of labour. Some socialists propose that these officials should owe their appointment to a central governing body; others propose that they should owe them to popular election; but in either case, appointment, promotion, or degradation would necessarily and avowedly, if it did not depend on favouritism, depend on the practical results which the different men in question elicited from labour by their different methods of directing it. In other words, the whole system of socialistic production would involve and depend on competition; and the only essential difference between this bureaucratic competition under socialism and the competition of capitalists which socialists so furiously denounce, is that whilst the capitalists obtain control over labour by means of wages, which control, by a natural and automatic process, is gradually extinguished unless it is used efficiently, the competitors for office under socialism would obtain the same control by compulsory powers with which the State would invest them, and which they would lose or retain at the pleasure of some more or less arbitrary authority.

Competition between employers, then, is a part of every social system that permits of progress,   Competition, then, between the directors of labour—or, as it is here defined, the struggle for industrial domination—is as much a part of the theoretical *régime* of socialism as it is a part of the actual *régime* of capitalism. The only differences between the two consist, firstly, in the means by which labour is directed, coercion being employed in one case, and in the other the inducement of wages ; and, secondly, in the means by which the fittest director is placed in power, and the less fit deprived of it—an official body deciding the matter in the one case, and the mass of the consuming public deciding it in the other for themselves.

but since the reintroduction of slavery is practically impossible, we must regard the wage-system   Now we may safely say that the *régime* of industrial coercion, or slavery, even though it should bear the name of socialism, is not in these days possible. It is impossible for two reasons—one, that it is out of harmony with the sentiments of the modern world ; and the other — equally strong, though not so generally avowed — that it is an exceedingly clumsy and wasteful instrument of

competition. We may, accordingly, dismiss it from
our consideration ; and such being the case, there
remains for us the absolute certainty that if society is as a permanent
to make any further industrial advance, or if it is to feature of
progressive
save itself from a relapse into industrial helplessness, societies.
the capitalistic wage-system, and with it capitalistic
competition, or, in other words, the competitive
struggle for domination, must both of them be continued under some form or other ; nor, although they
may be modified in an indefinite number of their details, is there any apparent possibility of ever modi-
fying them in any of their essentials. Indeed, the
great moral to be drawn from the facts that have been
here elucidated is that if any one institution in the
modern world threatens to be permanent, that institution is the capitalistic wage-system ; and all proposed
alterations in it we may set down as impossible in
precise proportion as the socialists attach value to
them. The foolish dreamers who imagine that they
can overthrow it, consider only its outer aspect, and
not the forces of which it is the expression. It is We might
reduce society
perfectly true that this system might at any given to ashes, but
time, and in any given country, be paralysed or this system
and capitalistic
reduced to ashes ; but the forces that would over- competition
would rise
throw it would be essentially non-productive. The again out of
them ;
men who destroyed it would find themselves powerless without it, and would be obliged to submit to,
and assist in, its reconstruction. For the outer form
of capitalism is not what capitalism is, any more
than a painter's brush is the power that paints
great pictures. Capitalism, in its essence, is merely

the realised process of the more efficient members of the human race controlling and guiding the less efficient; capitalistic competition is the means by
which, out of these more efficient members, society itself selects those who serve it best ; and no society which intends to remain civilised, and is not prepared to return to the direct coercion of slavery, can escape from competition and the wage-system, under some form or other, any more than it can stand in its own shadow.

With regard, then, to economic production, which, of all social activities, is for the practical sociologist incomparably the most important, what we have thus far seen is as follows. We have seen, not that it is impossible—for this question has been expressly postponed—that men may be made far more equal than they are now in respect of the possession of wealth ; but that whatever degree of equality they may some day attain to in its possession, they can never be otherwise than unequal in the parts played
by them in its production ; that their inequality in productive power is of such a kind as to render the industrial obedience of the larger number of them to the minority the primary and permanent condition on which economic progress is possible ; that what feather-brained fanatics call " *economic freedom* " would be merely another name for economic help-lessness ; and that all the democratic formulas which for the past hundred years have represented the ·employed as the producers of wealth, and the capitalistic employers as the appropriators of it, are,

instead of being, as they claim to be, the expressions
of a profound truth, related to truth only as being direct inversions of it. Whatever appearances may seem to show to the contrary, it is the few and not the many who, in the domain of economic ·production, are essentially and permanently the chief repositories of power. That this is so in the domain of intellect we have seen already. We will now turn our attention to the domain of political government, and consider the part played by the exceptional few there—the nature and origin of their power, and the means by which it is exercised.

CHAPTER IV·

In discussing the means by which the great man wields power in politics the debatable question differs from the question raised by his power in industry ; IN discussing, with reference to political government, the means by which the great man controls the actions of others, it will be found that the point on which we shall have to concentrate our attention differs somewhat from that which engaged it when we were discussing the same question with reference to economic production. For all the points which, with reference to the directors of industry, it was necessary to establish in opposition to the sociological sophistries of to-day are, with reference to the political governor, admitted by all alike. Thus we shall find on reflection that the extremest democratic reformer, no less than the aristocrat or the strict upholder of autocracy, admits, firstly, that satisfactory governors must be exceptional or great men ; secondly, that the fittest great men can be secured by competition only ; and, thirdly, that however they are appointed, and whatever may be the principles on which they govern, their orders must in every case be enforced by virtually the same

for the points that are debated in the case of the great wealth-producer are admitted by all in the case of the governor.

sanctions. The last of these three facts—namely, that the commands of the governor must be enforced by some system of restraint and punishment for the disobedient—is sufficiently plain to require no further notice; but the two others, obvious as they really are, are not perhaps generally realised, and it will be well to give a few words to them.

That the efficient governor, though he need not always be a genius, must in some respects, at all events, be a great or exceptional man, is of course admitted by the advocates of autocracy, aristocracy, or oligarchy. All that requires to be shown is that it is admitted also by the thinkers who are most opposed to them—by socialists and extreme democrats. This admission on their part is implied in the notorious importance attached by them to the machinery of popular election; for popular election is simply an elaborate means of expressing the opinion of the people that out of so many possible governors, this one or that one is endowed with greater capacity than the others. If the capacities of all were equal, or if exceptional capacity was not required, the *personnel* of the government might be chosen by casting lots. Next, as to the question of competition, it must be obvious to every one that the popular election of governors is not only an admission that some few men out of many are greater or more capable than the rest, but is also, on the part of the candidates for election themselves, competition in one of its intensest and most sharply accentuated forms.

12

Book II
Chapter 4

There is a
competitive
element even
in autocracies,

Competition, indeed, is implicit in every form of government. Were it absent in any, it would be absent in complete autocracies ; but even in these it is latent, and always ready to come into operation ; for the most absolute autocrat, if he happen to make his rule sufficiently odious to a sufficient number of his subjects, — "*postquam cerdonibus esse timendus cœperat*"—will, as history shows us, be assassinated or got rid of somehow, and some other candidate for power, probably an autocrat also, will be put in his place, and will either retain or lose it, according as experiment shows him to be a tolerable ruler, or the reverse. Here is political competition in its most rudimentary form ; but it is competition none the less ; and it generally involves a competition more advanced than itself; for the most absolute autocrat is obliged to govern through ministers ; and these rise and fall according as experiment shows them to be fitter or less fit for the accomplishment of their master's purposes. If, then, even the power of the autocrat rests ultimately on competition and practical experiment, much more does the power of government, under aristocratic and oligarchic constitutions. Oligarchies invariably aim at ruling through their strongest members ; and which are the strongest is shown by experimental competition

only ; whilst political democracy, under all its forms, is experimental competition open and undisguised. A Gladstone remains in power because, as his years of office succeed each other, he satisfies the majority by the manner in which he governs them ; and his

power is taken from him when the majority cease
to be satisfied, not only because they are of opinion that he governs badly, but because they are of opinion that a Disraeli will govern better. A democracy, in fact, and an oligarchy, so far as competition is concerned, differ merely in the way in which the competitors are admitted to the arena, and in the number and character of the jury which awards the prizes.

Since, then, with regard to the points just dealt All parties also agree that laws with — namely, the necessity for great men as must be enforced by pains governors, for the selection of the fittest of them and penalties. by competition, and for the use of coercion and punishment as a means of enforcing orders—there is no essential difference between the most extreme democracy and its opposites, in what does that practical or theoretical difference between them consist, by which most undoubtedly the former is distinguished from the latter? The only essential point of difference between them lies, not in their respective schemes or theories of the machinery of government, or of their methods of electing governors, but in their theory of the powers which election Democrats are peculiar only in communicates to those elected. An elected governor, their theory that the sole whether chosen from a large or a small class, is, greatness according to the aristocratic or oligarchic theory, required in their governors chosen because he is personally wiser than those is a perceptive and executive who elect him ; and it is theoretically his mission, greatness, which will within very wide limits, to follow his own judgment, enable them to carry out the not that of the electors. The democratic theory spontaneous is the very reverse of this. The elected governor, wishes of the many.

according to that theory, is elected not because he is supposed to be wiser than his constituents, but because he is supposed to be exceptionally capable of understanding their precise wishes, and giving effect to each of them. In the first of these two cases the governor is like the physician whom the patient calls in, but whose orders he never thinks of disputing. In the second, he is like the professional Spanish letter-writer, whom the illiterate lover employs to put his passion for him grammatically upon paper.

The only point, then, in which democracy can claim to differ essentially, not only from autocracy, but from any form of oligarchy, lies not in its form of government, but in the power that is behind its government. This power, according to democratic theorists, is the power of the mass of ordinary men, as definitely opposed to exceptional men; and the exceptional men who are picked out as governors would necessarily, in an ideal democracy, be exceptional only for such qualities as practical activity and a quick apprehension of the wishes of other people, which would enable them to do what their many-headed master bade them; but they would have to be wanting in any strength of mind or originality which might prompt them to acts out of harmony with their master's temper at the moment, or what is the same thing, to any acts beyond their master's comprehension, even although such acts might be for his future benefit. This is what the democratic theory, in its last analysis, means. All exceptional will is to be smothered or over-ridden by the average will, as is expressed clearly enough in the well-worn democratic formula—every man's vote is to count for one in government ; no man's vote is to count for more than one.

Now this theory of the relation of the great man to the many, so far as regards the conduct of civil government, is identical with the theory which, with a much wider application, Mr. Herbert Spencer enunciates as the foundation of his sociological system. As enunciated by Mr. Spencer we have already submitted it to examination, and we have shown that, in every practical sense, it is altogether fallacious, and that its acceptance renders all practical sociology impossible. We will now proceed to show that, as applied even to the most popular forms of government, it is as false as it is when applied to social phenomena generally.

That the essential principle of democracy, as just described, according to which the brain of the ideal ruler is merely a balance for weighing the wills of multitudes, which are dropped into one or other of its scales, like marbles—that this principle has ever yet been completely realised, no democrat will perhaps venture to maintain ; but the whole democratic propagandism of the present day implies, before all things else, that its complete realisation is possible, and that every day "the peoples" are getting nearer to it. The facts, however, which are supposed to warrant this conclusion are to be sought, not in the sphere of official government, but

The demo-
cratic ruler is,
theoretically,
a balance for
weighing the
wills of the
many ;

Book II
Chapter 4

without it. They are to be sought not in the conduct of elected legislators, but in the machinery by which they are elected, and, above all, in those unofficial movements, meetings, and agitations by which the prophets of democracy affirm that the great mass of the people is learning to exert the power which was always latent in it, and to express its will with regard to every question of government as it arises, even if it has something yet to learn in the art of securing that its governors shall carry out its commands.   It is this view of the situation which is expressed in the popular saying

or a machine for executing their "mandates";

that a constituency has elected a member, or that the people has elected a parliament, with what is called a "mandate" to do some specified thing or things—to break up the United Kingdom, to disestablish the English Church, to penalise the drinking of a glass of beer on Sundays, or to deprive our soldiers of protection against the most malignant of contagious maladies.

and there are signs which might suggest that the few in politics are really becoming the mere instruments of the many.

Now the democrats, it must be admitted, are so far right, that a real political power has come into existence which has no constitutional connection with the men who nominally govern; and this is frequently used with such efficiency, and with such definite purpose, that official governors—men of most exceptional intellect—are compelled by it to use their intellect for ends which they themselves condemn.   Here, then, in this external power, is to be found, if it is to be found anywhere, the will of the many, as conceived of by the theorists of demo-

cracy, exerting itself independently of any separate
will of the few, and turning the powers of the few
into its willing or unwilling instruments.

Now perhaps the question which will in this place
most naturally suggest itself is whether this will of
the many, however effectively it may be exercised,
is really a power that makes for civilisation and
progress, and whether it is not more likely to bring
harm than benefit to those very collections of
ordinary men who exercise it. And this question is,
no doubt, extremely pertinent; but it is not one that
need engage our attention now. The fact which
alone we are now concerned to demonstrate is that
the alleged will of the many is not what democrats
conceive it to be, and that it is not really the will of
the many at all.

For although there is much in the history of the But these signs
are deceptive ;
present century to warrant the assumption that the for what seems
the will of the
political will of the many is at last emerging as a many, really
supreme and independent governing power, we depends on the
action of
shall find that these movements and opinions, which another
minority.
seem, when viewed superficially, to result from the
spontaneous actions and spontaneous thoughts of
the many, really imply the influence of exceptional
men, just as much as those movements which are
avowedly aristocratic in origin ; and that in the
absence of these men the movements could never
have taken place, nor the opinions have ever
assumed any uniform and coherent shape.

To understand how this is, we need merely reflect
upon the fact that masses of men, as masses, can

184 ARISTOCRACY AND EVOLUTION

Opinions, to derive power from the numbers who hold them, must be identical ;

only have a will at all when their judgments with regard to certain particular questions happen to be absolutely identical, and have thus a cumulative force, like that of weights piled on one another above some substance which it is desired to compress.   Now, whatever may be the thoughts, wishes, or opinions which spontaneously shape themselves in the minds of any body of ordinary men—men various in training and temperament, and none of them remarkable for wisdom—these never take a shape which will give them any cumulative power unless amongst the ordinary men there is some man more active than the rest, who weighs them, compares them, eliminates what he thinks to be their discrepancies, adds what is in his opinion necessary to their logical completion, and clothes them in catching language, which appeals both to the mind and to the memory. Not till this is done do the mass of persons concerned

but they seldom are identical till a few men have manipulated them.

realise how identical their opinions on a given question are ; and they then perceive them to be identical for an exceedingly simple reason—that the exceptional man has made a mould for them, into which they have all been run.

It is then, for the first time, that the mass of ordinary men become conscious of corporate power ; for then they become, with regard to a given question, conscious for the first time that their opinions are absolutely identical, and that in a certain given direction their power is consequently cumulative. But the opinion of these men, whose numbers give political force to it, is very far from representing the capacities of these men only. It represents the capacities, the character, and very probably the personal designs of the exceptional man who supplied that common mould to which the unanimity of the other men's opinions is due; and the one opinion which thus comes to be held by all of them will not be precisely the opinion that was originally held by any. The original opinion of each will have undergone some modification. It will have been softened, emphasised, developed, or other elements will have been added to it, which would never have entered the mind of the ordinary man naturally, and which even when admitted he does but imperfectly understand. Thus whilst a political opinion expressed, or a political demand made, by a body of ordinary men thus absolutely unanimous seems at first sight a genuine expression of the will and the capacities of the many, it always in part, and it very often mainly represents capacities and purposes belonging to one man alone, the many being practically little more than a phonograph, which repeats his words to the world through an enormous resonator.

Thus what seems to be the opinion of the many is generally dependent on the influence of a few.

Let us take, for instance, the two questions of Free Trade and Bimetallism. If any British Government were to revert to the system of protection, it cannot be doubted that throughout the country there would be meetings and demonstrations, at which every throat would be unanimous in shouting condemnation of their conduct. America has witnessed a precisely similar outburst in favour of a proposal

The many, for instance, would never have had any opinions on Free Trade or Bimetallism if a few men had not worked on them.

to remonetise silver. The issues raised, however, both by the free traders and the bimetallists,
are of a kind so complicated that exceedingly few
people would be able even to describe their nature
clearly enough to satisfy the most lenient examiner
who should set them a paper in economics. The
majority of those who declared for bimetallism in
America had as little to do with forming their own
opinions as the little boys would have in a preparatory school who should shout their approval of
some new emendation made by one of their masters
· of a corrupt passage in Pindar ; nor does that
British opinion in favour of free trade principles
which has caused our Government to adopt them,
and would hinder or prevent their repudiation,
rest in the minds of the majority of those who
hold it, on any larger amount of original thought
or knowledge. Ninety-nine free traders out of
a hundred would never have been free traders at
all if it had not been for the oratory of Cobden.
The least-educated portion of the citizens of the
United States would never have howled themselves
hoarse over an intricate financial problem if it had
not been for the oratory and the singular activity of
Mr. Bryan. Indeed, what is oratory itself, which in
all democracies, from that of Athens downwards, has
been essential to the work of government, but an
embodied expression of the fact that the many are
powerless, unless here and there some thinker will
think for them, and give them opinions which may
form a mould or a nucleus for their own ? Even a

village meeting is never got together without the agency of some one who is slightly more efficient than the rest. He need not be wiser than they. He very frequently is not; but he has some gift or other which qualifies him for taking the lead. His temperament is more active, his words flow more freely, or he is hampered by less insight into his own ignorance or imbecility; and his opinions are the nucleus round which those of the rest form themselves, and which generally imparts to them something of its own character, as a vinegar plant does to the liquor in which it is immersed.

Without some such nuclei afforded to the many by the few, popular thought is nebulous, and popular will unborn. An exceptional few are essential even to those revolutionary movements which have the destruction of the power of the few for their object. It is impossible for the many to attack one set of superiors, except by submitting themselves to the leadership or dictatorship of another set; and although these last may to a certain extent represent the multitude, it is usually just as true that the multitude represent them. The multitude cannot even unite to influence those exceptional persons to whom is entrusted the official work of government without placing themselves under the influence of another set of exceptional persons; and thus the extremest democracy will be found, if we only look below the surface, to be neither more nor less than an oligarchy disguised. It is, no doubt, true that those who actually govern do in a certain

*Book II
Chapter 4*

Popular opinion requires exceptional men, as nuclei, round which to form itself.

sense derive their power from the many. They do so even in countries where the supreme governor is an autocrat. In countries with a popular constitution they derive their power from the many by an organised and conscious system; but even in the extremest democracies the average men can exercise their power only by constant processes of surrendering it into the hands of exceptional men. They

surrender it into the hands of the exceptional men for the simple and enduring reason that, with very few exceptions, which will be examined in another place, it comes into existence only in the very act of surrendering it; and the many accordingly place themselves in the hands of the few because, from the very constitution of human nature, they cannot avoid doing so.

We thus see that even in that sphere of political action in which, if anywhere, the many should be independent of the few, the many without the few would have no power at all.

The apologists of democracy, however, have another argument left them. They may contend that the exceptional men, who are necessary to the development of the collective powers of ordinary men, though each of them is constantly, with regard to particular questions, following his own devices rather than the instructions of the electorate, do on the whole, and in the long-run, substantially carry out the intentions and devices of those who are theoretically their masters; and that though they may do what their masters could never have thought of for themselves, yet they can never continue to do anything of which their masters do not actually approve. Now even were this representation of the case true, it would leave untouched that broad and fundamental truth on which it is the primary purpose of the present work to insist. It would leave untouched the truth that the great mass of human beings are helpless without the assistance of a minority more efficient than themselves. If ninety-nine average men, through the aid of a hundredth man who is exceptional, can develop and give effect to a collective will, which is altogether their own, and originates entirely with themselves, but if they can neither develop it nor give effect to it unless the hundredth man lent them his services, the power of this one man is as essential to the power of the ninety-nine, as it would be if the orders which he executes had been largely originated by himself ; just as a lens is essential to the photographer's camera though its function is solely to focalise, not to colour, the rays transmitted by it. Accordingly, even on the above hypothesis, the modern democratic formula, which makes each man count for one, and nobody count for more than one, would, if judged scientifically, be absolutely and fundamentally false ; for the power ascribed by it to the accumulated faculties of equals would be really the power of equals united with the power of a superior ; and the difference between the equals and the superior would be at once apparent from this—that if one of the equals were subtracted, the power of the whole

Even were this wholly true, the current formulas of democracy would be false, for unequal men would be essential to executing the wishes of equals.

hundred would be diminished by one ninety-ninth only; but if the one superior were subtracted, it would collapse altogether. Thus the presence of the superior, and the terms on which his services can be secured, would even in this case be subjects on which the sociologist would be bound to bestow the same attention as he bestows at present on the activities of the ordinary men; and unless he should do this, his conclusions would be wholly valueless.

Now in reality the few are never mere passive agents;   As a matter of fact, however, the hypothesis that the superior few are ever the mere passive agents which the democratic theory assumes them to be is false; and it is as a rule false in exact proportion to the difficulty and importance of the cases to which it is applied. The qualities which enable men to organise the opinions of others are usually qualities which endow them with strong opinions of their own; and in addition to their own opinions, these men, with their exceptional vigour, have usually their own purposes also; and the popular will, as put into execution by them, is always modified, and very often metamorphosed, by what they themselves add to or subtract from it. Still it must be admitted that, in spite of their dependence on the few, the many can, and do to a great extent, impress their own

but nevertheless the many do impress their will on them to a great extent.   genuine will—the will and wishes of the average man as distinct from the will and wishes of the man who is in any way exceptional—on the exceptional men to whom their power is surrendered. The acts of the governing few may never entirely represent

the will and wishes of the average man, when these
acts are considered as a whole ; but they may be
forced to embody, and they generally do embody, a
certain element of what average men wish and will ;
and their character as a whole is profoundly modified in consequence. The question then is simply a
question of degree. What is the extent—or rather The question is
what is the utmost possible extent—of this genuine ^{to what extent?}
power of the many to make the faculties of the
exceptional few their servants? Is it great or
small ?

The reader will perceive that when this question This intro-
is asked our inquiry is gradually taking a new <sup>duces us to a
new side of the</sup>
turn, and that having started with asserting the <sup>problem—the
extent of the</sup>
claims of the great man as the author and sustainer <sup>powers of the
many.</sup>
of both intellectual and economic progress, we are
led, when we come to consider him as an agent in
the domain of politics, to inquire into what is done
by the average man, as well as into what is done by
him. And the reason for this is that in the domain This is greater
of politics the many, so far as direct and intentional <sup>in politics than
in industry ;</sup>
influence is concerned, are actually capable of play-
ing a far larger part than they are in the domain
of speculation or of advanced economic production.
A statesman like Mr. Gladstone might, without
absurdity, maintain that he had a mandate from the
many to grant home-rule to Ireland ; but nobody
could pretend that any body of mechanics had given
Watt a mandate to invent the steam-engine, or
that any one gave Newton a mandate to discover
the law of gravitation. And yet the reflection will

probably force itself upon every reader that if the many play a part in politics which is commensurate with that of the few, they play a part in intellectual and economic progress also. It would be useless for the few to unfold their thoughts and their discoveries

to the many, if the many were not, in various degrees, capable of assimilating and responding to them. Still less could the great man of industry realise his progressive inventions, or carry out his extending schemes of business, if it were not that an indefinite number of ordinary men — those "serviceable animals," as Mr. John Morley calls them — were endowed with capacities that enabled them to carry out his bidding. What would Mahomet have done if he had not had followers? What would Columbus have done if he had not had seamen? The reader, accordingly, will inevitably be led to urge that in attributing to the great men of the world the results which we have attributed to them, our statements are unmeaning, unless they are accepted as incomplete, and are understood to imply more than they have actually expressed. If no progress of any kind could have taken place without the many, surely, it will be argued, the many must have had some share in producing it; and unless we can assert and discriminate precisely what this share is — what are the phenomena of progress which are due to the activity of ordinary men—it is meaningless to assert that most of them are due to the activity of exceptional men.

And the larger part of this argument is perfectly

true. In dealing with the activities of the few, we have taken those of the many for granted. This general assumption, however, though inevitable at the beginning of our inquiry, has been provisional only. To any scientific conception of what is done exclusively by the few, an equally scientific conception of what is done by the many is essential. We must measure the former by the latter, as we measure mountains by their respective heights above the sea-level. That such a discrimination between the work of these two bodies is possible may be doubted by some ; and accordingly before we actually proceed to undertake it, we will dispose of the arguments that will be, and actually have been, advanced in proof of its impracticability, and set forth the principles on which it must be, and obviously can be, made.

Book II
Chapter 4
We had to take it for granted at starting. We must now examine it.

BOOK III

CHAPTER I

IN the first chapter of his *Principles of Political Economy* Mill alludes to the question raised by certain thinkers, of "*whether nature gives more assistance to labour in one kind of industry than another*"; and he endeavours to show that the question is useless and unanswerable. In every industry, he says, there would be no product at all unless nature gave something and labour did something. Each is "*absolutely indispensable*," and the part played by each is consequently "*indefinite and incommensurable.*" "*When two conditions*," he proceeds, "*are equally necessary for producing the effect at all, it is unmeaning to say that so much of it is produced by one, and so much by the other; it is like attempting to decide which half of a pair of scissors has most to do in the act of cutting, or which of the factors five and six contributes most to the production of thirty.*" If this argument is applicable to nature and labour as agents in the

Mill declares that when two agencies are essential to producing an effect, their respective contributions to it cannot be discriminated.

production of commodities, it is equally applicable to the few and the many as agents in the production of social progress generally; and the crisp phrases and illustrations which Mill employs in formulating it, put in the clearest and most forcible manner possible the whole class of objections referred to at the close of the last Book.

Mill argues thus with special reference to land and labour;
Mill brings the argument forward with special reference to agriculture. Let us take, he says in effect, the products of any farm; and it is obviously absurd to inquire which produces most of it—the fields or the farm labourers. Now if all labour were equal, and if there were only one farm in the world, or if every acre of land, when the same labour was applied to it, yielded the same amount of produce, this would, no doubt, be true. The actual state of the case is, however, widely different. Acres vary

but he overlooks what in actual life is the main feature in the case.
very greatly in fertility; and if the produce of one— the least fertile—when cultivated by a given amount of labour, be symbolised by ten loaves, the produce of others, when cultivated by the same labour, will be symbolised by loaves to the number of twelve, fifteen, or twenty. Here, then, we have a constant quantity of labour, which produces ten loaves from each of the four acres in question; but when

The labour remaining the same, the product varies with the quality of the land.
applied to the first, it produces ten loaves only; when applied to the three others, it produces two, or five, or ten loaves in addition. About the first ten loaves, in each case, it is not possible to argue. So far as they are concerned, the result is in each case the same; with regard to them we cannot

make any comparison; and we must admit that the parts played by land and labour in producing them are *"indefinite and incommensurable,"* precisely as Mill says they are. But the two, the five, or the ten extra loaves which result when labour is applied to the second, the third, and the fourth acre. respectively, but do not result at all so long as it is applied only to the first, constitute phenomena of a different order altogether. The labour being in each of the four cases the same, and these additional loaves resulting in three cases only, these additional loaves are obviously not due to labour, but to certain additional qualities present in the last three acres and not present in the first. In other words, though in producing the loaves, or, as Mill puts it, *"the effect,"* the parts played respectively by land and labour are incommensurable so long as the land, the labour, and the effect remain the same, the parts become immediately mensurable as soon as the effect begins to vary, and one of the causes, and one of the causes only, varies also.

The extra product resulting from labour on superior land is due to the land, not to labour.

This truth can ˌbe yet further elucidated by means of Mill's two other illustrations. If the two blades of a pair of scissors were made of two different materials, and the one blade were of such a nature that it was always of the same quality, and human ingenuity was not capable of improving it, whilst the qualities of the other blade varied with the skill devoted to its manufacture, and if one pair of scissors should cut twenty yards of cloth in a minute, whilst another cut only ten, the additional

This is easily proved by a number of analogous illustrations.

efficiency of the more efficient pair would, it is perfectly obvious, be due to that blade in respect of which this pair differed from the pair which was less efficient, not to the blade in respect of which both pairs were similar.   Again, let us take Mill's case of the two numerals five and six.   If five is always to be the number multiplied, and six is always to be the multiplier, it is true we cannot say which does most in producing the result—thirty. But if the number to be multiplied remains always five, whilst the multiplying number varies—if it is in one case six and in another case ten,—and if the result of the multiplication in the second case is not thirty but fifty, it is obvious that the additional twenty which results from our multiplying by ten is due not to any change in the number multiplied, but to the additional four introduced into the number multiplying.   To these illustrations we may add two others—the movement of a modern bicycle and the movement of a man running.   A modern bicycle cannot be propelled without a chain ; and if there were only one kind of bicycle in the world, Mill might fairly have said that it was meaningless and useless to ask whether the wheels or the chain contributed most to its velocity.   But if there are two bicycles, with precisely similar wheels, but with dissimilar chains, and if the same man riding on one can accomplish ten miles an hour only, but on the other fifteen, the common sense of every bicycle rider in the world will tell him that the additional five miles are contributed entirely by the chain, and

the patentees of the chain, we may be certain,
will add their valuable testimony to the fact.   So
with regard to running, Mill might fairly have said
that if we consider it in an abstract and general
sense, it is absurd to ask which contributes most to
"*the effect*"—the ground or the man that runs on it,
because the first is as indispensable to the man's
movement as is the second.   But if two men are
racing each other over the same course, and one
runs a mile whilst the other runs only half, it is
perfectly obvious that the extra speed of the winner
is contributed not by the ground, which for both
men is just the same, but by certain qualities in the
winner which the loser does not possess, or which
the winner possesses in larger measure than he.

Now in all questions connected with progressive
social action the effects which have to be considered
are not general effects, such as running at some
indeterminate speed, each of which effects is considered as being single of its kind, and which, in
consequence, cannot be compared with anything,
but effects each kind of which exhibits many comparable varieties, such as the running of several
men whose respective speeds are different.   The
whole error of Mill's argument depends on his
failure to perceive this.   He describes the result
of man's labour applied to land—a result which we
have for convenience' sake expressed in terms of
loaves as "*the* effect."   He says "*nature and
labour are equally necessary for producing the effect
at all*," as though the same amount of land and

labour must always result in the production of the same number of loaves.   To conceive and speak of the matter in this way is to ignore entirely all the phenomena of progress—all the phenomena which differentiate civilisation from savagery, and which it is the special function of economics and of sociology to explain.   Rent, for example, the theory of which Mill states with extreme lucidity, and insists upon with the utmost emphasis, arises from the fact that one man and one acre of land,

The case of
labour directed
by different
great men is
the same as
the case of
labour applied
to different
qualities of
land.   The
great men pro-
duce the in-
crement. instead of producing something that can be described generally as "*the* effect," produce in different cases effects that are widely different—ten loaves when the acre is bad, twenty loaves when the acre is good : and, in a similar way, when the acres are of the same quality, twenty loaves will be produced by an acre if it is cultivated by the methods of civilisation, and only ten by an acre if it is cultivated by the methods of a savage.   Now, just as agricultural rent arises from different qualities in the soil, so does agricultural progress arise from differences in the powers of the men.   It is measured by, and it consists of, not "*the* effect," but a series of effects, similar indeed in kind, but continually increasing in degree ; and it is their differences in degree, not their similarity in kind, that form for the economist the particular subject to be considered.

   And what is true in this respect of production and progress in agriculture is equally true of pro-duction and progress generally.   The former indeed are the simplest type of the latter, just as they are their original basis ; and before we proceed farther, there is one fact more in connection with them on which it is necessary for the purposes of our present argument to insist. Of soils the same as to area, but not the same as to quality, some, it has been said, will produce ten loaves, some fifteen, some twenty ; and soils may exist, perhaps, which would produce only five. But in order that any soil may be cultivated by human labour, it is necessary that the product should be at least sufficient to keep the men alive who devote their labour to cultivating it. No set of men, unless artificially subsidised, could continue to cultivate any region if the product of twelve months' labour would support them for only three months. It follows, therefore, from this truism that no soils can be cultivated which will not yield to labour a certain minimum product. Now, though this minimum is, in a certain sense, the product of labour and of land jointly, for all purposes of practical reasoning it is the product of labour alone. It is so because the sole object of practical reasoning about the matter is to determine the principles on which the product of the land is to be distributed ; and with regard to that minimum there can be no doubt or question. It must go to the labourer, and it can go to no one else. The landlord, if there be one, cannot take any part of it ; for if he did, the labourer would die, and there would cease to be any product to take. Labour, then, in agriculture must be held for all practical purposes to produce the whole of that minimum resulting from its application to the least productive soils which the labourer can live by cultivating ; and it is only in the case of soils which are more productive than these, and which yield to similar labour a product above this minimum, that land,. apart from labour, can be said practically to produce anything at all.

Now just as we can argue with regard to land and labour, so can we argue with regard to the average men and the great men, and measure what they contribute respectively to any given civilisation; for just as a thousand men from some good soil will elicit twice the produce they would be able to elicit from a bad soil, so from a bad soil may a thousand average men manage to elicit, if directed by some agricultural genius, twice the product which they would elicit if left to themselves ; and just as in the former case, according to the principles above stated, we shall ascribe the smaller product to labour without any reference to land, and ascribe to land the excess only of the larger product over the smaller, so in the second shall we ascribe the smaller product to the average men, and the excess of the larger product over the smaller to the great man.

We shall say, in fact, that the great man produces so much of the product as comes annually into existence when he directs the others, and disappears as soon as he ceases to direct them.

Here, however, the original objection of Mill will suggest itself again, though in a somewhat different form ; for in spite of all that has been said, it still remains certain that the great man could not produce this excess unless the average men were present to carry out his directions; and the reader will possibly be disposed to argue that the average men may be as reasonably credited with the whole of the product except that insignificant fraction which the great man could have produced without *them*, as the great man may be credited with the whole of the product except that which the average men could have produced without *him*.

Labour, it is true, is essential to the production of the increment also;

Now this reasoning has a certain fanciful plausibility, but it is absolutely devoid of any practical meaning; and in order to show the reader how and why it is so, it will be necessary to direct his attention to a certain fact which lies at the bottom of all practical reasoning, but which few practical reasoners ever consciously realise. All such reasoning is in its nature hypothetical, and can be reduced to a statement that if such conditions are present, such consequences will result; and that if existing conditions be altered in any specified way, the results will exhibit a specified and corresponding difference. If, however, this reasoning is to have any practical value, one thing is essential to it—namely, that the supposed alterations shall be at least approximately possible. No practical conclusion, for instance, could possibly be drawn as to machinery by considering what would happen if the properties of the circle were to be changed, and different parts of the circumference should be at different distances from the

but we cannot draw any conclusions from the hypothesis of labour ceasing;

Book III
Chapter 1

for the
labourers
would have to
labour whether
the great men
were there or
no.

centre. It is equally evident that no practical
conclusion as to the claims and prospects of labour
could be drawn by considering what would happen
if the labourers could live without food. Now since
no food is producible without labour, a population
which does not labour is just as impossible a conception as a population which does not require to
eat ; and no practical conclusions can be arrived at
by supposing it to exist ; but populations which have
developed and submitted themselves to no great
men, not only can exist, but have existed, and do
exist to-day; and thus we are reasoning in a
strictly practical way when we consider what would
be produced by the average men if the great man
ceased to direct them, but we are reasoning to no
practical purpose at all by considering what would

The cessation
of the great
man's influence
is a practical
alternative ;
the cessation
of labour
is not ;

happen if the average men ceased to labour. The
latter—or the majority of them—would have to
labour in any case, whether there were any great
man to direct their labour or no; and the supposition
of their labouring is bound up with the supposition
of their existence. The sole practical alternatives
which can in the present case be conceived and
reasoned from are average men labouring under the
direction of the great man's talents, or the same men
labouring blindly as best they can by themselves.

These alternatives are being constantly exemplified in the actual life of communities. We may see
men to-day, not only amongst savages, but amongst
the peasantries of civilised countries, such as Russia,
India, and parts of Ireland and the Scottish islands,

who are still almost independent of any intellect superior to their own, and who maintain themselves by the exertion of man's commonest faculties only. We may see again populations who have been in the same condition, but who, under great men's guidance, become agents in producing a civilisation which they could by themselves not only not produce, but could, by themselves, hardly even imagine ; and again we may see how in more than one country the energies of the great man, having worked these wonders for a time, become paralysed by insecurity under a barbarous and predatory despotism, and how, as his action ceases, the masses relapse again into their former condition of relative inefficiency.

Accordingly, though the productivity of the average men, as distinct from the great men, will be different in one race or region from what it is in another, just as their diet will be and the other necessaries of existence, yet within each community experience furnishes us with comparisons which show us, roughly at all events, how much the average men produce without the aid of the great men, and how much the great men, by directing the average men, add to this.[1]   To calculate these amounts

---

[1] It is, of course, true that in densely populated countries and in certain industries the average workmen, if left to themselves suddenly, with no man of business ability to guide them, would be unable to produce anything.   But so long as the man of exceptional talent employs them to produce anything, they contribute something to the result, and must, for practical purposes, be held to produce so much of it as will provide them with the means of living.   If it happens, as is sometimes the case, that the total value of the profit

with any approach to exactness will, no doubt, be more difficult in some cases than others, just as is the case with book-keeping in various businesses. But it is enough to have shown the reader that, despite Mill's contention to the contrary, the calculation is one which is based on the simplest

Thus the great and most indisputable principles, and that not only
man, in the
most practical in a theoretical, but in the most strictly practical
sense, pro-
duces what sense, what great men produce, when they co-
labour would operate with average men by directing them, is
not produce in
his absence. the amount or degree in which the total result pro-
duced exceeds or excels that which was produced by average men when unaided, and would be again produced by them were the great man's aid withdrawn.

An analysis of   The absolute validity of this method of argu-
practical
reasoning as to ment and calculation will be yet more apparent
causes gener-
ally will show to the reader when we pursue a step farther
us the truth of our analysis of reasoning generally as applied
this.
to practical matters, and consider it especially when it takes the form of a direct discussion with regard to causes and effects. In the strictest sense of the word it would plainly be quite impossible to specify fully the causes of even effects of the simplest kind. The motion, for instance, of a ball when a cricketer hits it, would, in any discussion of the game, be said to have been caused by

is less than the workmen's wages, the employer must either alter the character of his product, so as to meet the public demand, or he will otherwise be crushed out of existence as an employer, and his workmen will pass under the control of some more able rival.

the cricketer; but the entire antecedents and conditions which have rendered this effect possible
comprise not only all the incidents of the cricketer's
past training, but the history of cricket itself, and
half the properties of matter. It would be impossible
and useless to specify all these. When we say that
anything is the cause of anything else, we are always
selecting that cause out of an indefinite number, on
which, for the purpose on hand, it is practically
important that we should insist ; and the cause on
which it is important that we should insist for For practical
practical purposes will be found to be always one purposes *the* cause of an
which, under the circumstances in view, may or effect is that cause only
may not be present,[1]—which a careless person may which may or may not be
neglect to introduce, or an ignorant person be present ;
persuaded to take away ; whilst those other causes
whose presence is assumed by all parties to the dis-

[1] It was his complete neglect of these considerations that enabled
Karl Marx to impose on himself and others his doctrine that the
value of commodities depended on the amount of average labour
embodied in them—a doctrine which is the most remarkable intellectual mare's nest of the century. It is perfectly true that if all
other circumstances were always equal—the demand for the commodities in question, the ability with which average labour is
directed, and the assistance which the genius of the great inventors
gives to it—it is perfectly true that then the amount of average labour
embodied in various commodities would be the measure of their
value ; for labour in that case would be the only variant. But, in
reality, the important variants are not average labour, but the ability
by which labour is directed. The efficiency of labour itself is
practically constant ; and for the student of wealth-production the
principal force to be studied is the ability of the few, by which the
labour of the many is multiplied, and which only exerts itself under
special social circumstances.

14

cussion, and which no one proposes to take away, or which no one is able to take away, or whose number, if they were mentioned, would make all discussion impossible, are passed over in silence, for there is no need to mention them. Thus we
as we see
when men
discuss the
cause of a fire,
all know that when a house is burnt to the ground the causes of the phenomenon comprise the inflammable nature of timber, and indeed the whole chemistry of combustion ; but if an insurance office is disputing the owner's claim to compensation on the ground that the owner set a light to it purposely, whilst the owner maintains that the scullerymaid set it alight by accident whilst reading in bed a novel of Belgravian life, the only causes that will be put forward by the litigants will, let us say, be a candle alleged by the owner to have ignited the scullery-maid's pillow-case accidentally, and on the other hand a match which is alleged by the agent of the insurance office to have been applied by the owner to the drawing-room curtains intentionally. Or again, let us take the case of a ship's chronometer.
or of the
accuracy of a
chronometer,
The reliability of a chronometer, any practical man will tell us if we ask him about the matter, depends on the balance and the escapement. It is the perfect "compensation" of the former and what is called the "detachment" of the latter that differentiates the chronometer from the ordinary lever watch ; and these are rightly said to be the causes of the chronometer's superiority as a time-keeper. But a balance and escapement of themselves will not keep time at all. They are useless without a main-

spring and a train of intervening wheel-work. But if any one were explaining the causes of a chronometer's exceptional accuracy he would never think of mentioning these last at all. He would not dwell on the properties of the coil of elastic steel, or on the interaction of the ordinary toothed wheels, or on the steel axes which make their interaction possible. And why would he omit these causes? He would omit them because they would be assumed, because there would be no discussion about them, and because they are implied in the existence of all watches and chronometers equally. If, however, the case were reversed—if all escapements and all balances were alike, and there was no room for superiority except in the main-spring and the wheel-work—the latter would be dwelt on, and the former would be passed over, in any discussion that turned on the causes of accurate time-keeping.

Let us take one case more. A man is hanging by or the causes of danger to a man who is hanging on to a rope which is fastened to a spike of rock, and is looking for samphire or birds' eggs on the face of a sheer cliff. It is suddenly perceived by some of his friends on the summit that the rope is frayed a yard or two above his head. They are anxious for his safety; and if any one asked them why, they would answer, Because his life depends on the rope not breaking. Let us suppose, however, that the rope is perfectly strong, but that the spike of rock it is attached to shows signs of being about to fall. The man's friends in that case will explain

Book III
Chapter I

their anxiety by saying that his life depends not on the rope but on the rock. In either case it would literally depend on both, and on a thousand other things beside ; but in either case one cause only is mentioned, or calls for mention, and that is the cause whose cessation or continuance is doubtful. For similar reasons, and in a similar sense, great men are said to be the causes of all that is done or produced in the communities to which they belong, beyond a certain minimum which, even when not insignificant, is stationary.; for though the efforts of the average men are essential to the production of this addition to the minimum, just as they are to the production of the minimum itself, there is no question of their efforts coming to an end unless the men come to an end also; whereas the activities of the great men require special circumstances for their development, and constitute the only productive force which modern democratic activity practically tends to paralyse, or at all events diminish or impede.

But there is
another means
of discriminat-
ing between the
products of
exceptional
men and
ordinary men.

But there is yet another method, still more necessary to be described, by which we are able to differentiate the respective products of these two classes of men—a method which will assist us not only to assign.to each a certain portion of one joint effect, but also to particularise many of the elements of which each portion is composed. This method will be explained more fully in the following chapter, but it will be well to give a general and preliminary explanation of it here. It is founded on the two following propositions, which, when once they have been considered, will be seen to be self-evident. Whatever the many contribute to the social conditions of a community, either in the way of industrial production or of the formation of habits and sentiments, consists of effects produced by those traits or faculties of human nature in which all members of that community are approximately and practically equal. Thus the fact that all men are alike obliged to eat, and that all parents as a rule have a preference for their own offspring, are facts which determine much in the conditions of all societies. On the other hand the social effects which are produced exclusively by the few are effects produced by certain traits and faculties which, though possibly possessed in a rudimentary state by all men, are appreciably and efficiently developed in the persons of the few only. The dramas of Shakespeare, though in a sense they are eminently national, could never have been produced had Shakespeare possessed no gifts except such as were possessed at the time by the English nation at large. The discoveries of Newton, the inventions of Watt and Stephenson, similarly were produced by powers that were indefinitely above the average. It is needless to say that they could not have been produced otherwise. If we will but reflect carefully on obvious truths like these, we shall see that civilisations are woven out of two kinds of materials, the one originating in traits common to the community generally, the other in traits confined to a

Marginal notes:

Book III
Chapter 1

This is by an analysis of the faculties necessary to produce the product.

Are these faculties possessed by all, or by a few only?

more or less numerous minority; and even when
the two are most closely woven together we shall
be able to follow out and identify the different
threads, which never can lose the trace of their
different and opposite origins.

CHAPTER II

THE NATURE AND SCOPE OF PURELY DEMOCRATIC ACTION, OR THE ACTION OF AVERAGE MEN IN CO-OPERATION.

THE great-man theory as held by the conventional historian, and expressed by Carlyle and others in those vehement formulas which have so justly excited the ridicule of Mr. Herbert Spencer, errs not because it emphasises the fact that the great man is the sole cause of progress in the sense that no progress could have taken place without him, but because it ignores the fact that the ordinary men of his time, being the tools with which he works, or the instrument on which he plays, the result is conditioned not only by his capacities, but by theirs; just as the kind of music that can be produced by a pianist is determined not only by his own skill, but by the character of the piano also. Writers like Mr. Spencer, on the other hand, and with him the whole school of socialists, impressed by the obvious fact that the many do something, never pause to inquire what they do, or how much they do, or how little, but rush to the conclusion that

Carlyle was wrong in his claims for the great man because he failed to note that his powers were conditioned by the capacities of the ordinary men influenced by him.

The socialists are wrong because, seeing that the many do something, they argue that they do everything.

the many do everything. This conclusion is even more meaningless than the doctrine which it is intended to contradict. The many do something, and they do what is of extreme importance ; but its importance is strictly limited, and is indeed only intelligible through its limitations, just as the character of a profile is intelligible only through its outlines. The object, therefore, of the sociological inquirer must be to discover precisely what these limitations are. The methods by which the discovery is to be made have been already indicated. Let us now go on to apply them. They are of two kinds. One consists of an examination of what, in any domain of activity, the many would produce, if the influence of the few were absent. The other consists in an examination of the kind of faculties which the production of such or such a result implies. If these faculties are common to all, we say the result is produced by the many ; if the faculties are rare, we say it is produced by the few.

What the many do is limited. We must see precisely what the limits are.

If a Russian conspirator employs a hundred workmen to dig what they think is a cellar,

The practical validity of both these kinds of reasoning is shown by the following imaginary, but not impossible case. A hundred Russian workmen, all of them loyal to the Czar, are employed by a citizen of Moscow to enlarge a subterranean cellar, and another hundred are employed to fill it with heavy wine-cases. A week after the work is completed the Czar is driving outside, and, as he passes the citizen's house, is killed by an explosion

but is a mine for blowing up the Czar,

from below. The so-called cellar was a mine, the wine-cases were filled with dynamite. Now if all

those who were concerned in the production of this
catastrophe were tried, it is perfectly evident that the part played by the workmen would be sharply separated from that played by the man employing them ; and that, though they no doubt would have contributed something to the result, they would have contributed nothing to its essential and criminal elements. It is equally evident that if the designed and attained result had been not criminal, but beneficent, the elements in it that made it glorious would be the product of the man who planned and intended it, and not of the workmen who blindly obeyed his orders, neither knowing nor caring what the result would be. Let us take another case of a somewhat different character. When a spontaneous cheer bursts from a thousand people, the volume of sound is obviously the unadulterated product of the many. On the other hand, when a thousand people with ordinarily good voices are so trained and organised as to sing a chorus out of *Israel in Egypt*, the peculiar qualities which render the sounds produced by them valuable, obviously imply the existence of the musical genius of Handel, or in other words, faculties which belong to hardly one man in a million, and are thus the product not of the many, but of one.

the conspirator contributes the entire criminal character of the enterprise.

When a choir sing Handel's music, Handel contributes the specific character of the sounds sung by them.

And now let us turn to the actual facts of life, and the kinds of activity on which progress and civilisation depend, and let us apply our two analytical methods to these. It is needless to repeat, after what has been said in a previous chapter, that it is

Let us turn to the facts of social progress,

impossible, in a case like this, to examine social
activity as a whole. Such activity is of various kinds,
and each must be dealt with separately. Let us
begin, then, with two—the activity of economic production, and the activity which results in the growth
and begin with of speculative knowledge. The first affords us the
economic
progress and clearest illustration of how to discriminate the pro-
progress in
knowledge. duct of the many by considering what it would
shrink to were the influence of the few absent. The
second affords us the clearest illustration of how to
discriminate the product of the many by considering
the nature of the faculties which the production of
the result implies.

In the case of   To begin with production, then, let us take the
economic
progress we case of the United Kingdom, and consider the amount
must apply the per head that was annually produced by the popula-
method of
inquiring what tion a hundred years ago. This amount was about
is produced by
labour with £14. At the present time it is something like
and without
the assistance £35, and the purchasing power of money has so
of the great
man. increased with the cheapening of commodities, that
the excess of the latter· sum over the former is far
greater than it seems. Now, if we attribute the
entire production of this country, at the close of the
last century, to common or average labour (which is
plainly an absurd concession), we shall gain some
idea of what the utmost limits of the independent
productivity of the ordinary man are ; for the
ordinary man's talents as a producer, when directed
by nobody but himself, have, as has been said
already, not appreciably increased in the course of
two thousand years, and have certainly not increased

within the past three generations. The only thing
that has increased has been the concentration on the ordinary man's productive talents of the productive talents of the exceptional man. The talents of the exceptional man, in fact, have been the only variant in the problem ; and, accordingly, the minimum which these talents produce is the total difference between £14 and £35. This sum is no mere piece of fanciful ingenuity. Parts of it are being done daily before our eyes, and its practical character is being shown in the most conclusive manner, when the profits of a business decline on the death of some head or partner, or when some declining town is restored to its old prosperity by some man of industrial genius, who starts in it some new manufacture.

And now let us pass from industrial activity to intellectual, and apply to this our second method of analysis. Of purely intellectual results, or, as Mill calls them, "*advances in speculative knowledge*," the most striking examples are to be found in the mathematical sciences. To the advances made in these it is not only certain but obvious, that the many have contributed nothing, because even of that section of mankind which has some mathematical aptitude the majority are unable even to appreciate them completely when they are made ; much less do they possess the powers to make them. No one would contend that the books of Euclid are the result of the faculties possessed by every average school-boy, or of the kind of man into which the average school-boy grows. We may indeed dismiss purely intellectual progress as the domain in which the efficiency of the many stands absolutely at zero.

Let us pass now to the domain of political government, and consider to what extent the faculties of the many, as distinct from those of the few, are capable of operating there. This inquiry resolves itself mainly into the question of how much the many can do to direct the activity of the few, the activity of the few being presupposed; but it will be well to consider first how much, if anything, the many can accomplish, or the faculties of ordinary men can accomplish, without any assistance from exceptional faculties whatsoever. In the domain of politics, which is here meant to include all organised action of a public and political character, as well as the making and the administration of laws, the only positive functions or actions which can be performed by the co-operation of the average faculties of men, or by absolute and unadulterated democracy, are very simple destructive actions and the formulation of, and the insistence on, very simple demands. Of the destructive actions referred to we shall find an excellent example in the lynching of a negro who has outraged some white American girl, or in such an act as the burning of the Tuileries by the communists. In each of these actions the feelings of those who take part in it are as nearly as possible identical. In the first, all of the men are equal in their sense of righteous indignation; in the second, they are all equal in their feeling of blind rebellion; and no special skill is in either case

required by any one of them. It is true that even
in such cases as these there will most probably be
leaders, of some sort, but they will be leaders by
accident, and the others will be their comrades
rather than their subordinates. Of the simple
demands which the many can formulate and insist and formulate
upon unaided, we may take as an example a demand only the
simplest
for the abolition of a tax which distresses in an demands.
obvious way multitudes of men equally ; or a
demand for the continuance of a war, in which the
issues at stake are sufficiently apparent to anybody
who can read a newspaper. The protest against
the tax by the multitudes of men whom it harasses,
and the national demand, when it arises, for the
continuance of such a war, are phenomena which
are absolutely democratic. They are each the sum
of a number of spontaneous feelings and reasonings.
They do not require any leader to stimulate them ;
and all who contribute to their force do so in an
equal degree.

But the moment we come to cases of any com- The moment
matters
plexity the situation changes. If the negro's guilt become at all
complex, the
could be established only by inference, the lynchers faculties of the
exceptional
would have to be convinced of it by some clever man are
advocate. If the lynching itself were a matter of required.
extreme difficulty, the lynchers would require to be
commanded by the boldest and shrewdest of their
number. If the tax protested against were indirect,
if its injurious effects were hard to detect and
realise, and if it were capable of being represented as
less injurious than any other, men of exceptional

activity and exceptional sharpness would be required to rouse the sufferers to a perception of what caused their suffering. In other words, democracy, the many, or the faculties possessed by the many, are incapable of initiation in any complex matter, or of carrying out any course of complex action when initiated; and we may sum up the case by saying that all corporate action in politics is less and less purely democratic in proportion as the questions dealt with are less and less simple.

Now in any civilised country few governmental measures are really simple.
Now, as a matter of fact, in any civilised country the majority of the measures which the Government has to devise and carry out, however simple in appearance, are very far from simple in reality. Even when their details are few, the good or the bad effects of them are certain to depend on a great variety of circumstances, with regard to which ordinary faculties can form no independent judgment; and if ordinary men are to express any judgment on such measures at all which is not put into their mouths by others and then uttered by rote, these measures must be placed before them by

Exceptional men must simplify them for the many.
talented interpreters and advocates, who will reduce the details to a real or apparent simplicity and invest their alleged results with charm and an air of certainty.[1] Accordingly, when we approach the

---

[1] This truth is strikingly illustrated by the history of the Home Rule agitation in Ireland. Whether Home Rule would be advantageous for the British Empire or for Ireland is a very complicated question, and the demand for it consequently never became genuinely popular until it was identified with the simplest of all aspirations—the non-payment of rent.

question from the point of view of the many, we do
nothing but arrive at the same conclusion to which we were brought when we approached it from the point of view of the few. We arrive, that is to say, at the conclusion that, if we mean by government the devising, the passing, and the administration of this and of that measure, the genuine power of the many, even under the most popular constitution, becomes less and less in proportion as the greatness and the civilisation of the country increase. The voice of the many is heard as loudly as ever ; but what guides the voice is not the personality that seems to utter it. What guides it is a handful of men, exceptionally active, though not always exceptionally wise. The voice is the voice of Jacob, but the hands are the hands of Esau.

And here before pursuing the subject farther let us look back for a moment, and consider the point in our argument at which we have now arrived. We have seen, then, that in the domain of modern industrial activity the many, if we estimate the total produced in terms of value, produce only an insignificant portion of the total. We have seen that in the domain of intellectual and speculative progress the many literally produce or achieve nothing. We have seen that in the devising and administration of governmental measures the many are powerful in proportion as the issues are exceptionally simple —that is to say, in proportion as they are few and far between.

Now the reader may think that this brings us to

the end of our inquiry; but it only brings us to the beginning of what is really the important part of it.

For though these conclusions, so far as they go, are absolutely true, they by no means dispose of the whole question which is before us, nor do they really reduce the social power of the many to such small dimensions as they at first sight seem to do. Thus speculative knowledge, though the many contribute nothing to its progress, itself contributes nothing to progress until the many are affected by it, and respond somehow to its stimulus; economic production, when regarded merely as an affair of quantity or as an accumulation of values—a process in which the part played by the many is humble— does not represent that process in its true social entirety; nor is civil government wholly an affair of measures which are devised, discussed, amended, demanded, opposed, carried, or rejected from year to year. We shall find, accordingly, that, in spite of what has just been said, there is room in social life for the operation of the genuine will of the many— of pure, spontaneous, and unadulterated democracy. We shall find that the power of this will, though it is in certain directions incalculably less than it is at present generally believed to be, is paramount in domains where its action is not generally recognised at all; and the nature of its action here will throw a remarkable light on the nature of all action which is in a true sense democratic. Of the domains of activity here referred to, the most important are those of religion and family life.

Every religion, regarded as a body of doctrines
and observances, with the special habits of mind
and dispositions of the heart which are appropriate
to them, which has ever influenced great masses of
mankind, is mainly a result of pure democratic action.
It is true that in the establishment of the great
religions of the world another agency has played a
great part also.   In no other sphere has the influence
of great individuals been so vast and so far-reaching
as in this.   The mere mention of such personages as
Christ, Buddha, and Mahomet will make us realise
that such is the case ; and to these we may add the
missionaries, saints, and theologians who have spread
and explained the respective gospels entrusted to
them, and given by their saintly lives examples of
the value of their teaching.   But whilst nowhere is
the power of the few—of the very few—more
conspicuous than in the domain of religion, nowhere
is the power of the many more conspicuous also.
No religion has ever grown, become established, and
influenced the lives of men unless its doctrines and
its spirit have appealed to those wants of the heart
and soul which have been shared, to a degree
approximately equal, by all members of the communities, nations, or races amongst whom the religion
in question has become established.

The truth of this statement is not in the least
invalidated if we apply it to a religion which we assume to have been supernaturally revealed.   Indeed,
the clearest example of its truth may be found in the
phenomenon of Christianity.   Whether we attri-

Book III
Chapter 2

Though the
influence of
the great man
in religion is
enormous,

yet religions
have only
grown and
endured be-
cause they
touch the heart
of the average
man.

Christianity
exemplifies
this fact ;

bute the doctrines of Christianity to a natural or a supernatural source, it will be equally plain in either case that they have found acceptance amongst men because there was something inherent in the nature of each individual Christian which naturally responded to them. Even the staunchest Protestant who takes his stand most exclusively on the Bible will be unable to deny that Protestant Christianity, as it exists, represents not merely an assent to a number of bare propositions uttered by Christ, or made with regard to Him by His disciples, but also the subjective interpretation given to these by each believer as he assents to them. Thus the doctrine of the Atonement would never have been accepted by men, it would never even have conveyed any meaning to them, if there had not been something in their nature corresponding to a sense of sin ; and the universal effect which, for a time at least, this doctrine had on all the Western nations and on all classes alike, showed that this something which corresponded with the sense of sin was one of those characteristics in which all men were approximately equal, and that the acceptance of the doctrine was therefore a true act of democracy.

and especially
Catholicism.
But the clearest illustration of the truth thus insisted on is to be found, not amongst the varying and conflicting doctrines of Protestantism, which represent theoretically the direct result of the revealed truths of the Bible on each believer individually, but in Christianity as represented by the Church of Rome. According to ordinary Protestant opinion,

the doctrines of the Church of Rome represent a structure built up by the misguided ingenuity of priests, and imposed by them on a credulous and passive laity; but so far, at all events, as the more important doctrines are concerned, the very reverse is the case really. It has been the world of ordinary believers that has imposed its beliefs on the priests; not the priests that have imposed them on the world of ordinary believers. Let us take, for instance, the Catholic doctrine of the Eucharist, or the beliefs implied in the *cultus* of the Virgin Mary. That the sacramental elements were actually the body and blood of Christ, that the Redeemer who died on the cross for each individual sinner entered under the form of these elements into each sinner's body—entered bearing the stripes on it by which the sinner was healed, and mixing with the sinner's blood the divine blood that had been shed for him—this was the belief of the common unlettered communicant long before priests and theologians had, by the aid of Aristotle, explained the assumed miracle as a process of transubstantiation; and longer still before their philosophic explanation was, by the ratification of any general Council, given its place amongst the definite teachings of the Church. Similarly, the devotion to the Virgin Mary first sprang up amongst the mass of believers naturally, because the idea of God's mother, with all her motherly love, with all her virgin purity, and with all her human sorrows allied so closely to omnipotence, touched countless hearts

Book III

Chapter 2

The doctrines formulated by the aristocracy of Popes and Councils originated amongst the mass of common believers.

in a way which was in all cases practically similar ; just as the offer of a helping hand would make a similar appeal to each one of a multitude of men drowning.   The official teaching of the Church with

Theologians
and Councils
merely
reasoned on
the materials
thus given
them. regard to the Virgin's sinlessness, and the degree of worship which is her due, has been the work, no doubt, of the few, not of the many—of priests, of theologians, of Councils, of the spiritual aristocracy ; but the doctrines which they have thus defined have been no more fabricated by themselves than the wines, good or bad, which a peasantry have made for centuries, are made by the chemist of to-day, who at last undertakes to analyse them.

Catholicism
shows the great
part played by
the many so
clearly because
the part played
by the few is
defined by it
so sharply.   It has been said that the part which democracy plays in the development of religion is shown us by the Church of Rome with greater distinctness than it is by any other great communion of believers ; and the reason is that no other great communion of believers shows us with so much precision the part played by an aristocracy, and thus leaves the part played by democracy with so sharply defined a frontier.   The Roman Church alone is in possession of a complete machinery by which all the pious opinions of the whole body of its members—the opinions which have spontaneously shaped them- selves in the minds of innumerable Christians as the result of a multitude of independent spiritual experi- ences, and which, when sufficiently manifested, have been studied by various theologians, and reduced by them to logical and coherent forms—shall be finally submitted to one great representative

Council. This Council considers how far they are   Book III
consistent with doctrines already defined, and with   Chapter 2
one another, and how far, explicitly or implicitly,
there is any warrant for them in the Scriptures. It
ends with rejecting some, whilst others are reconciled
and affirmed by it; and then these last are added
to the authoritative teachings of the Church. But
the Council, with the Pope included in it, is nothing
more than a lens by which the rays originating in
the democracy of the faithful are focalised and made
to transmit a clear and coherent picture ; and the
Roman Catholic religion, regarded as a body of
doctrines which have actually influenced the spiritual
lives of men, is a magnified picture, projected, as
it were, upon the sky, of those secret but common
elements of the human mind and heart, in virtue of
which all men are supposed to be equal before God,
and which unite the faithful into one class, instead
of graduating them into many.

This analysis of what may be called the natural   Catholicism,
however, is
history of Catholicism may be thought, perhaps,   only alluded to
here because it
to have little appreciable connection with those   illustrates the
social or sociological problems which at present   essential nature
of truly demo-
agitate the world, and give to the theory of de-   cratic action.
mocracy its main practical interest. But neither
Catholicism nor religion at large has been referred
to here for its own sake. They have been referred
to because the case of religion affords a singularly
clear illustration of the essential nature of democratic
action generally, because it helps us to understand
that action in the affairs of ordinary life, and

Book III
Chapter 2

because it shows us very vividly how democracy, as a political power, operates outside the domain to which it is popularly supposed to be confined.[1]

Thus en-
lightened by it,
let us turn
back to family
life.

And now let us turn again to a nation's family life, and consider it in the light which the case of Catholicism throws on the question of what, essentially, democratic action is. The religious life of a Catholic is meritorious only when the beliefs and dispositions of heart which his religion requires of him are spontaneous. No doubt they may have been developed in him by some stimulus from without, but it is essential that, when once present in him, they should draw their life from himself. A saint may rouse a sinner to repentance, but the repentance in its minutest details must be the sinner's own work. He must be his own overseer, he must be his own taskmaster. In economic production this is not so. A bricklayer may contribute to the building of some exquisite cathedral without any sympathy with the architect's intentions, and indeed without any knowledge of them ; but a man cannot be a true Christian unless Christ's will becomes his, and unless the beliefs suggested from without are seized on by his own soul, and made a part of himself by his soul's spon-

[1] The political power of the religious beliefs of a community can be seen at a glance when we consider our own government of India. Our government there, in the ordinary sense of the word, is a government of the few, not a government of the many ; and yet the religion or religions of the many impose limitations on our legislators as stringent as any that could be imposed on them by any number of formal mandates.

taneous workings. Thus the common religious opinions of the mass of devout Catholics are, theoretically at all events, the sum of a number of independent opinions, which agree because they result from a number of similar but independent experiences. Here we have the essence of democratic action— namely, a natural coincidence of conclusions, which happen to be identical, not because those who hold them have allowed their thinking to be done for them by the same thinkers, but because with regard to the points in question they naturally themselves think and feel identically.

Book III
Chapter 2

Catholicism shows that democracy is a natural coincidence of conclusions.

Now the home or family lives of the citizens of any race or nation owe their points of identity to essentially the same causes. They result from propensities in a vast multitude of men which, although they are similar, are independent. The structure of the family differs amongst different races. Amongst some it is based on polygamy; amongst others on monogamy; but no matter what its details in either case may be, the government, however autocratic, accommodates itself to the family life of the people; not the family life of the people to the laws and the dictation of the government. It will be enough to confine ourselves to the Western or progressive races, amongst whom family life has its basis in monogamy. Advocates of socialism often distinctly say, and the principles of socialism beyond all doubt require, that the family, as now existing, shall be practically broken up; and that whilst the union of the parents is

The home life of a race depends on the same co-incidence, or on spontaneously similar propensities.

made terminable with an ease unapproached at present, the multiplication of children shall be regulated by State authority, and that the children themselves shall be reared by the State rather than by the parents. For both these arrangements there are many obvious arguments, which are from the point of view of the socialist quite unanswerable. If the State binds itself to provide for all the children that are born, it is bound to claim some control over the number of them that shall be thrown on its hands. If the State is to be the sole employer and sole director of labour, it must settle the number of children that shall be educated for each branch of industry. If the solidarity of feeling requisite to make socialism possible is ever to be obtained, it can be obtained only by fusing into one those family groups now so obstinately separate. But here the socialists encounter one of their great stumbling-blocks.[1] In theory the advocates of the extremest and most complete democracy, they are baffled by the habits and character of the very masses to whom they address themselves. There may be unhappy homes, and there may be unnatural parents, but the masses, as a whole, will not listen to any proposal for invading the privacy of the home or for tampering with the parental tie. Any average

[1] The Italian socialist, Giovanni Rossi, who attempted in 1890 to found a socialistic colony in Brazil (an attempt which completely failed), attributes his failure largely to the tenacity with which his followers clung to family life. " If I had the power," he writes, " to banish the greatest afflictions of this word, plagues, wars, famines, etc. etc., I would renounce it, if instead I could suppress the family."

mother would, when it came to the point, tear out
the eyes of any socialist legislator who, under
pretext of increasing her weekly wages, should
seriously attempt to snatch her children out of her
arms. Similar resistance would be offered to any
attempt to modify, beyond certain limits, the institution of marriage, or to interfere in any way with the
habits of a people's home life. These habits give This truly
rise to legislation by the few, but they do not democratic coincidence
originate in it. The legislation of the few, on the forces all governments
contrary, has so to shape itself as to protect those to accommo-date themselves
modes of life and institutions which these habits to it.
naturally produce; and the laws that do this, no
matter who devises and administers them, come into
being under genuinely democratic dictation. It is
a genuinely democratic power which maintains them
unaltered, or imposes its own limits on any modification of them which may be made.

The effects, however, of the natural similarities The same
of men's family lives are not to be found only in democratic power deter-
the domain of laws and government. They confront mines the structure of our
us even more openly in the material surroundings of houses,
our existence, especially in the structure of the
dwellings of all classes except the lowest. The
detached cottage as well as the large mansion, the
row of cottages each with its separate door, and the
tenement of three rooms, are in one respect all alike.
They are constructed and arranged in accordance
with those propensities which keep the members of
the family group united, and each family group
separate from all others. Nor do matters end here;

Book III Chapter 2

for if the propensities which result in family life affect the structure of the dwelling, other tastes or propensities equally spontaneous determine what commodities shall be put in it. . It is true that these tastes are different in different social classes; it is true

and the furniture and other commodities in them;

also that they have not, so far as their details are concerned, as deep a root in our nature as the propensities which give its character to the family. They are stimulated, sustained, and modified by constant suggestions from without, by circumstances, and by tastes which, within limits, vary greatly ; but they are all alike in this, that when they become efficient, or, in other words, take definite shape as a want, the want has become a part of the man who feels it, and is for the time as spontaneous as are the family instincts themselves.

and indeed on all economic products.

The influence, however, of men's spontaneous wants is not confined to the house and household appliances, but extends itself over the whole domain of economic products. And here we are brought back again to another portion of the ground which we have already traversed. We are brought back to the domain of economic production ; but brought back with eyes opened to a new order of facts.

Now before we proceed to a consideration of these, let us recapitulate what has been said with regard to this subject already. The main fact which was dwelt upon in our previous examination of it was the fact that in wealth-production all but the earlier advances are due, both in their achievement and their maintenance, to the few,

and to the few alone.   The practical validity of this
reasoning has been shown in the preceding chapter,
and defended against the common objections sure to
be brought against it ; and just now it was reinforced
incidentally when we were considering the influence
of the many on the doctrines of the Church of Rome ;
for whilst the essentially democratic origin of these
doctrines was insisted on, it was shown that the
religion of the Catholic democracy could have no
organic growth, no definition nor cohesion without
the aristocracy of theologians and the machinery of
popes and councils.   It was further pointed out For though in
that if even in the development of religion the production
many are dependent on the exceptional powers of dependent on
the few, in the process of economic production they the few,
are incalculably more dependent.   For whilst
Catholicism represents the ideas of the multitude,
analysed, perfected, and carried out by the few,
advanced economic production, such as the production of a beautiful cathedral, represents the ideas of
the few carried out in partial or complete ignorance
by the multitude.

Attention must now be called to certain further
facts which constitute the final evidence of the truth
of the same conclusions.

The facts now referred to are those of con- (a fact which
temporary trade unionism.   These are supposed trade unionism
by many of the trade unionists and their sympathisers more
to show the growth of democratic power in the apparent)
domain of production generally.   What they do in
reality is to exhibit its essential limitations.   They

show this in a way which is hidden from the careless thinker by a curiously inaccurate and misleading use of language. Trade unionism is constantly described as the organisation of Labour. In reality it is nothing of the kind. It is an organisation of labourers ; and that, as we shall see, is a totally different thing ; for where labourers are spoken of under the collective name of Labour, they are so spoken of with special and exclusive reference to the phenomena which they manifest when actually exerting themselves in production. Were the same men organised for some ethical or religious purpose, they would be spoken of not as Labour, but as the National or Popular Conscience. The organisation of Labour is the setting men to perform a large variety of correlated productive tasks, and prescribing to each man what his own task shall be. But the organisation of labourers that has been brought about by trade unionism is of a precisely opposite kind, and has a precisely opposite end. Its end is not production, but the cessation of production ; not the prescribing, the devising, and the allotting of tasks, but the taking men away from them. In a word, it is the organisation not of production, but of obstruction ; nor does the fact that the trade unions have succeeded in organising the latter give so much as a hint that they would be able to organise the former. Even if they could do so, it would be the leaders, not the men, that performed the feat — a new race of employers separating themselves from the body of the em-ployed ; and this fact is oddly enough acknowledged by the very men who are apparently most blind to it. For one of the arguments most frequently used to show the practicability of industrial democracy is based on the unusual ability manifested by the officials of the trade unions in managing strikes and great demonstrations of strikers. Must not these men, it is asked, have very exceptional capacities who can gather together their thousands at the shortest possible notice, and march them into Hyde Park through the crowded thoroughfares of London ? And it is perfectly true that many of the trade union leaders are, in their own way, men with remarkable and exceptional characteristics. But, in the first place, the more that their admirers magnify them, the more do they detract from the democratic character of trade unionism ; and in the second place, if a man is necessarily exceptional because he can so far organise some thousands of men as to march them occasionally into an enclosure where they walk about sucking oranges, how much more exceptional must be the abilities that can organise similar men, day after day, for the per- formance of the most intricately adjusted tasks, in such a way that their efforts shall result in an Atlantic liner ! Trade unionism, then, whatever the ability of its leaders, does not represent democratic action, in the actual process of economic production at all ; and instead of pointing to any development of such action in the future, merely helps to show us that no such development is to be looked for.

Such being the case, then, the facts that now

claim our attention will, when they are first stated, wear an appearance of paradox; for though the power of democracy, in the advanced processes of production, is smaller than it is in any other kind of social activity, abstract thought and discovery alone excepted, yet it exercises an influence on production none the less, which is as purely democratic in character and as far-reaching in its consequences as that which it has ever exercised over the doctrines of any religion.

yet it is the wants and tastes of the many which determine what shall be produced;
For what is the object of production ?   It is the satisfaction of human wants, which begin as needs, and gradually develop into tastes.   The multiplication of these needs, together with the satisfaction of them, is what civilisation means; and though material wealth may increase, as it does in many new countries, without any concurrent development of civilisation in its higher forms, civilisation in its higher forms cannot increase, and certainly cannot diffuse itself throughout the community at large, without a development in the means of material production.   Books, for example, though they are vehicles of mental culture, are themselves economic commodities, and depend for their accessibility to the public on the same kind of industrial agencies as do cotton, sugar, tobacco, and that comforter of the nations—alcohol.   Refinement of taste and feeling, again, is largely diffused by pictures; but the accessibility of any great picture to the vast majority of any nation depends on the industrial processes by which it can be cheaply and faithfully reproduced—

processes which have only of late years reached any sort of perfection.

But all the industrial ingenuity that great men have ever possessed would be absolutely futile unless the commodities they were employed in producing, or the services they were employed in rendering, satisfied tastes and wants existing in various sections of the community. The eliciting of these wants, or the development of these tastes, depends often on the previous supply of the products or services that minister to them. Thus the introduction of railways, of the electric telegraph, of the telephone, of the electric light, preceded any popular demand for them ; and many a great writer, according to the well-known saying, has to create the taste by which he is to be appreciated. But he could not create the taste, or, in other words, make it actual, unless it existed already in human nature as a potentiality, any more than the producers of electric light could make the general public anxious to have it in their houses if mankind at large entertained no wish whatever to do anything but sleep between the hours of sunset and sunrise. The wants and tastes, then, to which all production ministers, whether common to all men, like the desire for food, or developed by influences from without, like the desire for telegraphic accommodation, are, when once they are in existence, essentially democratic in their nature. They are not like the movements of a mason, who constructs under an architect's order a cathedral with the design of which he has nothing at

and though great men elicit these wants by first supplying them,

the wants themselves must be latent in the nature of the many, and when once aroused are essentially democratic phenomena.

all to do.   They represent the uncontrolled promptings of the individual's own nature, and they affect
production, and dictate to the producers what they
shall produce, because they represent a spontaneous
similarity of taste amongst a multitude of·individuals
living under similar circumstances.   Here we have
the reconciliation of the seemingly contradictory facts,
that the power of the. many over production is at
once paramount and small.

Thus, though
economic sup-
ply is aristo-
cratic, eco-
nomic demand
is purely
democratic. Economic demand, though it owes most of its
development to the few, is yet, when its development has taken place, fundamentally democratic
in its nature.   But, on the other hand, economic
supply, which not only ministers to existing wants,
but elicits new ones, tends ever more and more as
civilisation advances to depend on the action of the
few.   For as wants increase there is required, in
order to satisfy them, a growing elaboration in the
methods and organisation of supply ; and in proportion as supply becomes more and more elaborately organised, it becomes, from the necessities of
the case, less and less democratic.   In the Middle
Ages, for instance, the only rich *supplying* class
consisted of merchants, because the exchange of
commodities, and the bringing them in the required
quantities to the proper markets, was a process
more complicated than the orginal processes of
producing them.   Production has now become quite
as complicated as commerce ; and a manufacturing
aristocracy has developed itself equal in wealth to
the commercial.

But though supply thus depends on the domina-
tion of the few, and rises and falls with the ability with which that domination is exercised, it is itself at the same time under the domina- tion of the many. Some industrial genius may make a colossal fortune by directing the labour of some thousands of men to the production (let us say) of a new species of beer ; but his enterprise will succeed only because millions of men like the beer, and demand it under the direction of their own taste alone. The tastes of the many, of course, exhibit many varieties. Where a million men demand beer, another million will demand whisky ; and there are many commodities, such as guns, golf balls, and cricket bats, the demand for which is confined to comparatively small classes. But the point here insisted on is, not that every member of the community demands the same commodities, but that whatever commodities are demanded, are demanded in each case in accordance with the· spontaneous wishes of individuals, and that the total force of the demand is the cumulative result of a number of actions and desires which happen to be spontaneously similar. The commodities supplied to them have, in other words, to be accommodated to a genuinely democratic order ; and if the consum- ing democracy does not consider them suitable, it virtually, by refusing to buy them, condemns them to be destroyed. Thus if we direct our attention to consumption, the few—the directors of industry— are the servants of the many ; though if we direct

our attention, as we did previously, to production, the many, in the capacity of workers, are the servants or subjects of the few.

And now let us turn back to the domain of politics. We shall find that we do so possessed of a new clue to the true nature and extent of the powers of the many there. For we shall find that in civil government, just as in economic production, the process involved is a process of supply and demand ; and that whilst there is a certain kind of political demand in respect of which the many are paramount, and act as a true democracy, their power in the business of supply is never more than partial, and is in most cases illusory.

The first point of which we must here take notice is this—that though the analogy between economic production and civil government is a genuine one, it is not to be found in the phenomena in which we should naturally be tempted to look for it. What we should naturally be inclined to do would be to take the demand for laws and policies as the counterpart to the demand for commodities, and the framing of such laws and the carrying out of policies as the counterpart to economic supply ; the first of these, like the demand for commodities, being simple and spontaneous ; the second difficult, like the manufacture of them. But in arguing thus we should be wrong.

The demand for laws and policies is, as we have seen already, by no means a simple thing, like the demand, let us say, for a particular kind of beer ; nor is it the true counterpart to such a demand ; for the beer is demanded for its own sake, but laws and policies are not. They are demanded for the sake of certain results on social life which, by various processes of reasoning, those who demand them have been led to believe that they will produce ; and it is the results of laws and policies, not the laws and policies themselves, which are in the political sphere what commodities are in the economic, and for which alone the demand is purely and genuinely democratic. The multitudes of men who were led to demand the abolition of the corn laws were not led to do so because the actual process of abolishing them was profitable or pleasurable in itself, but because they believed it would mean a larger loaf on their breakfast-tables. It was in the demand for the loaf that the many were spontaneously unanimous, and expressed their own views, not those of anybody else. Their unanimity in demanding the measure was produced by the arguments of an intellectual oligarchy, and could not have been produced without them. Thus whilst the demand for the larger loaf was equivalent to a demand for a particular kind of beer, the demand for the law was equivalent to a demand that the brewer should employ some novel appliances for brewing, with the merits of which they were acquainted only through the puffs and explanations of the patentee.

There is therefore a great difference ·between political demand and economic. Economic demand

The demand for laws is not the counter-part of a demand for commodities, for commodities are demanded for their own sake, laws for the sake of their results.

The demand for laws is like a demand that commodities should be made by some special kind of machinery.

Book III
Chapter 2
is single ; political demand is double ; and whilst one part of political demand—namely, the demand

No one makes
this latter
demand.
Economic de-
mand is single;
political
demand is
double.
for social results—corresponds with economic demand, or the demand of the consumer for commodities, the other part of political demand—namely, the demand for particular measures—does not correspond with economic demand at all, but is, on the contrary, in contrast to it.   For when workmen's wives buy some particular make of calico for their husband's shirts, or when cyclists buy some particular kind of tyre for their bicycles, they do so because they approve of the qualities which those goods manifest when in use ; not because they approve of the machinery by

Political demo-
cracy is vul-
garly identified
with the de-
mand, not for
social goods,
but for
machinery.
which the goods were made.   But in politics, although there is likewise a demand for political goods, as such,—for social security, personal prosperity, and so forth,—of which each man is naturally his own judge, just as those who use them are of the tyres or calico, and although statesmen and governments are frequently supported by the nation, not because they have carried this measure or that, but because the political goods supplied by them are on the whole satisfactory, yet the political demand which is supposed to be the special characteristic of democracies is not a demand for the completed goods, but a demand that this or that patent shall be used in the hope of producing them.

Now political patents are most of them highly complicated devices ; the action of all of them is dependent on a complication of circumstances ; and they are always the work of a special class of inventors. They never represent the spontaneously similar ideas of the mass of ordinary men, any more than the machinery used in a great brewery represents the spontaneously similar ideas of the happy and united customers whom a spontaneously similar taste leads to the same tied house. All that the many can do with regard to these political patents is to listen to the accounts of them given by the patentees, their agents, and their travellers, and to make the best choice they can between a number of different contrivances which they have had no share in devising, and which they only partially understand. They are, indeed, in much the same position in which that portion of the public would be placed which travels habitually between London and Glasgow, if it were asked to decide by its votes which of five kinds of reversing gear should be made use of on the London and North-Western engines. If this question had really to be decided by vote, the public might so far instruct itself by lectures from the competing inventors as to give votes for this contrivance or for that; but the very grounds on which its choice was formed would be obviously supplied to it by others; its choice would be limited by the number of the contrivances before it, and the part spontaneously played by it in the whole transaction would be small. And yet, as has just been said, it is the making of a choice of this kind that is regarded as being, in the domain of politics, typically, if not exclusively, the exercise of

Book III
Chapter 2

But in so far as democracy is a demand not for goods but for machinery, it is not purely democratic.

The demands of the many are manipulated by the few.

the power of the many. The result is that, whilst the many do in reality exert, through their spontaneously similar demand for certain social results, an influence on legislation which in certain respects is paramount, the political theorist, neglecting this fact altogether, confines himself to asserting their

Why, then, is democracy specially associated with the demand in which its power is least? power in the demand for political means—the kind of demand in respect of which they are most influenced by others.

Now what, let us ask, is the explanation of this fact? How does it come that in government a power is attributed to the many which is, even by recent socialists, not attributed to them in economic production? The reason is that over the processes of economic production the many can exercise no control at all, but that over the devising of governmental measures they can exercise some, which, though absolutely small, is yet, by comparison, large.

Because it is the only sphere of activity in which the many can interfere with the machinery of supply at all; Thus, for instance, though the structure and manufacture of watches is in one sense determined by the many, because the manufacture of those watches only can be continued permanently which satisfy the many, and which the many will consent to buy, it would be impossible for any watchmaker to produce good watches at all if his workmen were constantly required to be altering or readjusting the escapements in order to introduce some "dodge" devised by any man in the street. But in politics this is not the case. The influence of the men in the street, though it can exert itself through

exceptional men only, and is consequently not
wholly their own, does continually make itself felt in law-making as it does not make itself felt in watchmaking ; and yet the conduct of government is not rendered impossible, whereas the making of the watches would be. Indeed, in very many cases it is not even rendered unsatisfactory.

For this peculiarity in politics there are three reasons. One is that the connection between measures and the general welfare of the community is by no means so close or immediate as the connection between a watchmaker's tool and the wheel or pinion to which he applies it. Social effects follow on measures slowly, and the tendencies of bad measures are neutralised by other causes. The second reason is that, as Mr. Spencer rightly insists—agreeing in this judgment with the wisdom of Dr. Johnson—the social ills which governments "can cause or cure" are far less numerous than many thinkers imagine ; and the third reason is one with which we are already familiar, that the power of the many in determining what measures shall be adopted is, although not an illusion, less considerable than it appears to be. But whatever their power in this respect, the great point to remember is that it cannot exert itself or exist for any practical purpose unless the few provide it with the means of doing so, any more than a rudder has power to guide a ship unless some other power shall have set the ship in motion. The popular demand for measures, or the popular

and they can interfere with it here because the effects of political government on life are less close and less important than the effects of business management on business ;

and in any case the apparent power of the many is even here controlled by the few.

choice between them, alike presupposes the few who will make the supply a possibility.

And if the power of the many over supply is thus limited even in the domain of politics, in the domain of economic production it is more limited still, and in the domain of intellectual progress it is absolutely non-existent. Their true power is in their demand for completed results—for knowledge which they can assimilate, for dogmas logically stated, which reveal to them clearly what they already believe dimly, for food they can enjoy, for clothes that please their eyes, for commodities and appliances that minister to their comfort and con- venience, for social security, for freedom, and for personal and national prosperity. In other words, the truth, when properly understood, is a truism. The many are all powerful in determining the quality of progress and civilisation because it is their own tastes and wants to which civilisation must minister, and their own qualities which civilisation must draw out ; but of initiating civilisation, of advancing it, or even maintaining it, the many are absolutely in- capable unless they have the few to guide them. They contain within themselves the things that have to be developed, but they cannot themselves provide themselves with the conditions of their own development. Without the few to assist them they could no more progress than a train of railway carriages could progress in the absence of the locomotive.

It is impossible, however, to state these conclusions plainly without realising that in some quarters violent objections will be taken to them ; nor is it difficult to see on what grounds the objections will rest.   These shall accordingly be discussed in the next chapter ; and it shall be shown that the conclusions to which our inquiry has brought us thus far really contain in them nothing inconsistent with the sentiments, or incompatible with the objects, of even those extreme reformers who will certainly feel impelled to attack them.

CHAPTER III

THE QUALITIES OF THE ORDINARY, AS OPPOSED TO
THE GREAT, MAN

It will be objected that the conclusions reached in the last chapter derogate from the dignity of the average man.

THE objections which will be taken to the conclusion arrived at in the preceding chapter resolve themselves into two groups, one of which rests on general and more or less sentimental considerations, the other on practical. We will deal with the former first.

This group of objections will, by those persons who entertain them, be probably first expressed in an outburst of fine indignation at the wrong which the conclusions just epitomised do to the average man; for such persons will at once take them as implying that the average man is a miserable and helpless creature, with only enough intelligence to carry out blindly the orders which his betters are condescending enough to give him; and this implication will strike them as a wanton insult. They will think over various men in private and humble life who were never thought by themselves or others to be above the average level, but who yet were gifted with intelligence, taste, and skill equal to any possessed by the men who are called great. They will reflect that these men represent not the few, but the many ; and they will angrily reject a theory which frankly denies to the many any of those forces which specifically make for progress.

But this class of objections, which was already briefly glanced at when we were considering the precise points by which the great man is distinguished from the average man, will disappear altogether when we take the matter conversely and consider the precise points in which the average man differs from the great man.

In any discussion that aims at scientific precision it is necessary to give to the principal terms used a far more definite meaning than is given to them when they are used ordinarily ; for most words when used ordinarily have several meanings, but when used technically they must have only one. Any term, then, when used technically will of necessity specifically exclude a number of ideas—and it may be very important ones—which are frequently attached to it when it is used in conversation or general literature. This observation, as the reader will readily perceive, has a special application to our use of the term *great man*. The greatness of the great man, regarded as an agent of progress, is a quality, as has been said, which is to be measured by its overt results ; and its overt results consist of, and are brought about by, not what he does in his own person, but what he makes others do. It is needless to insist

Book III
Chapter 3

But they do not really do so ;

for since the great man, as here technically defined, is the man who influences others so as to promote progress, upon this truth again, as it has been explained at great length already, and it is impossible that any reader can misunderstand it. What it is necessary for us here to explain and insist upon is its converse

the ordinary man, as opposed to him, need not be stupid.
—namely, that if the essence of technical greatness is so to influence the actions or thoughts of other men that the productivity of human labour is increased or the scope of human thought enlarged, no man is technically great who is not in this way influential.

When we come to reflect closely on this definition, some of the results will strike us as not a little curious ; for if we exclude from the class of great

He is merely the man whose talents do not increase the efficiency of other men.
men and relegate to the class of ordinary men all those whose greatness begins and ends with themselves, and does not tend to communicate itself to any one beside themselves, so as to make others think or act more efficiently than they would unaided, ordinary men, or the many, in our present technical sense of the words, will include a number of men of the most brilliant capacities and accomplishments.

Poets, in this technical sense, are ordinary men.
The greatest poets, for instance, will in this way be classed as ordinary men, whilst the inventor of machinery for making good boots cheaply will be classed as a great man. And the reason is as follows. A great inventor is great as an agent of progress because when the apparatus invented by him is in process of being manufactured, and a thousand workmen are shaping or multiplying its separate parts, or again, when ten thousand other workmen are using the machines when completed, he makes each workman do precisely what he would

do himself if he were performing their several tasks
actually with his own hands. But a great poet—let
us say Shakespeare—could not in a similar way
so influence a thousand ordinary writers that they
should all of them be producing plays like *Macbeth*
or *Hamlet.* Indeed, the greater the poet is, the
more absolutely incommunicable is his gift. Shakespeare may have so far contributed to progress as to
have aided in the development of literary English
generally, but he has not, in the course of some three
hundred years, brought into existence one dramatist
comparable to himself.[1]   In art, in fact, after a
certain point has been passed, it can hardly be said
that there is any progress at all.

It is still more important to observe that what is
true of the arts is also true of the crafts, or, in other
words, those kinds of manual work whose special characteristic is rare personal skill. Manual skill, though
essential to material progress no less than unskilled
labour is, does not, except during the earlier stages of
civilisation, itself constitute an actively progressive
principle. That is to say, at a very early stage
in the development of productive industry manual
skill reaches its utmost limits, and thenceforward remains stationary, whilst industry continues to progress. Thus the skill which is evidenced by the

---

[1] Of course the great poet, like the great religious teacher, may
have an effect on the thoughts and imaginations of his readers, and
he may be a great man or an agent of progress in this way. But he
is not, in the technical sense of the word, a great man in reference to
his own art. He does not promote progress amongst other poets.

gem-engraving of the Greeks and Romans has rarely been equalled since, and has certainly never been surpassed. But we need not stop short at the antiquity of the Greeks and Romans. Many of the implements made by the prehistoric lake-dwellers could not, so far as mere manual workmanship is concerned, be better made by any workman or mechanic of to-day. Indeed, so far is the progress of material civilisation from depending on or coin-
for very great
manual skill
does not pro-
mote progress
or influence
others, ciding with any progress in manual skill, that it actually depends on a getting rid of the necessity, not certainly of all skill, but of skill of the rarer kinds. If any machine, for example, depended for its successful operation on an accurate finish in certain essential parts which only one workman in half a million could give, such a machine would be practically almost worthless. A productive machine is of use in the service of society generally in proportion as the machines or processes by which it is itself manufactured obviate the necessity for any skill in manufacturing it beyond such as can be obtained with considerable ease and constancy.

Many sentimentalists—and it is difficult not to sympathise with them—regret the manner in which manufacture is thus superseding craftsmanship, or that kind of production in which the beauty or excellence of the product is the direct result and expression of the skill of one producer. But this natural regret, though most frequently expressed by socialists, is defensible only on grounds of the narrowest social exclusiveness. That the artist-

craftsman who gives his talents directly to each particular commodity in the production of which he is
concerned—a silver cup, or a lamp, or a curiously-
designed carpet, or a printed volume—will produce
objects having a charm which is wanting in similar
objects produced by the methods of the manufacturer
is, no doubt, true.   But great artist-craftsmen being
few in number, the beautiful objects they make by
the craftsman's methods are few in number also, and
are consequently obtainable by a few persons only ;
whilst the objects inferior, but approximately similar
to them, which the great manufacturer multiplies in
indefinite quantities, are accessible to the many,
who, under any social system, must either have
these or have nothing of the kind at all.   An artist-
craftsman, for example, such as the late Mr.
William Morris, or a transcriber and illuminator in
a mediæval monastery, could produce a volume
indefinitely more beautiful than any product of the
steam printing-press ; but a book which the methods
of the manufacturer would admit of being sold for
sixpence might cost, if produced by the craftsman,
twice that number of pounds ; and it is easy to see
that, supposing a study of the Bible to be desirable,
a village comprising four hundred and eighty families
would be benefited more by each family having a
sixpenny Bible of its own than it would by the existence of one sumptuous copy chained to a desk in the
village church or reading-room.

Rare manual skill, in short, does not promote
progress, or help to maintain civilisation at any

Book III
Chapter 3

unless it can
be metamor-
phosed into
the shape of
orders given to
others.

given level, unless it can metamorphose itself—
as in many cases it can do by means of patterns
or otherwise—into a series of orders which men
who have less skill can execute, and thus affects
commodities not directly, but indirectly. So long
as it resides in exertions of the craftsman's hand,
applied directly to each commodity produced, it has
on the progress of the arts generally no effect at all.
The man or men who invented the slide rest communicated a new power to every one of the innumerable artisans now using it; but an artisan
who should produce exceptionally accurate work
owing to the exceptional accuracy and steadiness of
his own hand, could no more add anything to the
faculties of even one of his fellows than a beautiful
woman can, by means of her own beauty, improve
the eyes, nose, or hair of her plainer sisters.
Material progress, then, as has just been said, is
so far from being dependent on the growth of rare
manual skill that it takes place in proportion as the
necessity for such skill is eliminated.

Again, brilli-
ance or charm
in private life
does not pro-
mote progress.

And now let us turn from the consideration of
human capacities, as applied to and expressing
themselves in the production of particular commodities or results, and consider them as they
reveal themselves in ordinary life and conversation.
We shall find ourselves confronted by a similar set
of facts here. We shall see that many of the talents
and qualities which, when possessed by our friends
or by ourselves, elicit our strongest admiration, and
give an interest to human nature, do nothing to

advance or to maintain civilisation at all. No one,
for example, who knows anything of English society will deny that conversational wit is one of the rarest faculties to be met with in it, and earns for its possessor the reputation of an exceptionally brilliant man ; but its possession by one man does not cause its existence in others. The wit leaves the rest of society precisely where he found it. The same is the case with private goodness and wisdom. They may indeed affect an exceedingly small circle, but there is in their influence nothing certain or lasting. The most highly moral parents have often the most dissipated sons ; it requires almost as much wisdom to take sound advice as to give it ; even if the sensible and the excellent exert a good influence on their own friends, they have no tendency to inaugurate any general moral advance ; and a man whose life is rendered interesting by an exceptionally romantic passion may illustrate the capacities of human nature, but he does nothing to expand them.

It will thus be seen that when we describe the Therefore the majority of mankind as being so far passive with ordinary men, who do not regard to the production of progress that unless promote progress, are not there were a minority of men with faculties which asserted to be lacking in high the majority do not possess, no progress or civilisa- qualities. tion would take place at all, we are not declaring that the larger part of mankind are stupid, foolish, unskilful, or void of resource, or that human nature as exemplified in the normal man or woman is not often noble and beautiful, and is not always interesting. On the contrary, the very reverse is the

17

case. What is really interesting in human life and in human nature is the universal and typical elements in it, not the exceptional ; and we can show ourselves the truth of this in a very convincing way by looking into the mirror that is held up to nature by art.

The most famous and interesting characters to be found in fiction or in the drama, though they may have been invested by their creators with exceptional circumstances and endowed with exceptional gifts, have interested and appealed both to the world and their creators through the qualities and experiences which they share with human beings generally, not through those which may incidentally make them peculiar. Very few men, for example, are as intellectual as Hamlet ; but Hamlet has interested the world because, as has been well said of him, he is not " a man," but " man." If a great dramatist or novelist makes his heroes exceptional, he does so only because he can, by this device, more easily give a magnified representation of what is universal ; and the universal elements which he magnifies excite universal interest, not because they are exhibited on more than a common scale, but because they are thus exhibited with a more than common clearness. What are the most beautiful love-poems that have made their writers immortal but an expression of what is felt by millions, though it can be expressed only by a few ? Why is there life still in the two marriage songs of Catullus, if it were not for the living strings in the normal human heart which the magic of his hand still touches ?

But not only is the normal man the type of what is interesting and important in humanity. He is also the type of wise conduct in life, and secures amongst men in general a conformity to this conduct, not by means of advice given by exceptionally excellent individuals, but by the purely democratic pressure of cumulative class opinion. The force which this opinion exercises is commonly called " The World." The details of its injunctions and prohibitions are different in different classes ; and when it is called "The World," reference is usually being made to the pressure exercised by it in the highest classes only. But this limitation of meaning is altogether arbitrary. Every class is " The World," so far as regards itself. It has its own standards of manners, honour, prudence, dress, and also of moral judgment as applied to social conduct ; and it is in respect of all of them incalculably wiser than most individuals who differ from it. In social life even the greatest genius is ridiculous, in so far as he is unusual in anything except his greatness.

It is, moreover, the same cumulative common sense, the same spontaneous identity of perception on the part of ordinary men, that forms, as Aristotle says, the fundamental test of what is real. The world of reality is distinguished from the world of dreams because the former is the same for all men. It is ὁ πᾶσι δοκεῖ. The same fact is the foundation and the justification of trial by jury—an institution in which, as Sir Henry Maine has observed, we

*Book III
Chapter 3*

Average opinion also on social matters is for each class the wise opinion :

and the average faculties shared by all are in one sense the test of truth.

Book III
Chapter 3

Therefore in
denying to
average men
the powers
that produce
progress,

have the very abstract and essence of all practicable democratic government.

It is true that even here we are brought sharply back again to those limitations by which the powers of the normal man are surrounded. The jury, who represent the normal man's intelligence, require, as Sir Henry Maine points out, to have the facts on which they are to base their judgment, in exact proportion as these are obscure or complicated, reduced to order for them by advocates whose powers are more than normal. It is also true that, though it is the identity of ordinary men's perceptions which shows the reality and the qualities of external objects, ordinary men's perceptions would never have sufficed to show us that the earth was not the centre of the universe, and that the sun did not move round it. But the true moral of all that has been just insisted on is, that in denying to the masses of mankind those special powers which actively initiate and actively promote progress, and actively sustain the fabric of advanced civilisation,

we are not
degrading the
average man.
We are merely
asserting that
these powers
form but a
small part of
life.

we are not denying to the masses of mankind great moral and great intellectual qualities generally. We are not asserting that the normal, the average, the ordinary man is incapable of being developed into a creature endowed with beliefs, thoughts, and feelings which are not only noble and correct, but which expand and improve as civilisation advances. We are merely asserting that the ordinary man, or the masses of mankind, which are simply the ordinary man multiplied, cannot provide themselves with the conditions of their own progressive develop- ment ; or, to put the matter in a still more compre- hensive way, we are merely asserting that that particular form of greatness which improves those conditions or sustains them, by influencing, or com- pelling, or enabling masses of men to act or think as they would not act or think otherwise, consti- tutes a very small portion of human activity, and a still smaller portion of human life.

This truth has been lost sight of because modern social philosophers, led astray by political and other passions, have confused two distinct things—man as a moral being, moving in a circle of prescribed duties, and man as a being capable of public or social initiative ; and the more we study the ordinary man, and the more fully we appreciate the varied possibilities of his nature, the more clearly shall we see, and the more ungrudgingly shall we re- cognise, how absolutely he is, so far as civilisation is concerned, dependent on the exceptional man for even those very powers in virtue of which the action of the exceptional man is controlled by him.

The general or the sentimental objections, then, which might not unnaturally arise in the minds of many when the claims of the great man to be the sole agent of progress are first broadly asserted, are found to disappear altogether when the meaning of these claims is more fully considered. But senti- mental objections, as has been said already, are by no means the only objections which these claims have to encounter. Objections will be raised against them which are economic rather than sentimental, and which, moreover,—this is a still more important fact — rest solely upon a practical, and have no theoretical basis.

In order to see what these objections are it will be well to consider them in their extremest and most uncompromising form. We will accordingly consider them as put forward by the socialists. That the objections of the socialists to the claims made for the great man are not grounded in any theory that consistently disallows them, is sufficiently shown by the fact that even the most extreme socialists, no less than the members of every other militant party, are always extolling the exceptional qualities of their own leaders. Agitators, thinkers, and writers like Karl Marx, Lassalle, and Engels have been extolled by their followers as though in their own way equal to Cæsar and Napoleon, to Aristotle, Galileo, and Bacon; and their works are continually called "marvels of reasoning," and described as evincing "such powers of thought as are given to only a few men in the course of five hundred years." The arguments, therefore, which are employed by socialistic thinkers to convince them that the great man is not essential to social progress, and plays no real part in it—those arguments to the examination of which the first chapters of this work were devoted, do not really convince even those who lay most stress on them, so far as they are applicable to social progress generally. For the

socialists in practice are forced to limit the application of them to two kinds of social action only; and
these are social activity in the domains of political
government and of wealth-production. They are,
moreover, applied to the latter of these with so much
more strictness than to the former, that the objections to the special claims of the great man as a
wealth-producer are the only ones that here require
our attention.

   Now even here we shall find that the objections
in question are originated not by theoretical, but by
practical considerations only; for one of the most
curious features in the history of socialistic thought,
from the time when socialists claim that it first
began to be scientific till to-day, has been the unwilling replacement, in their theory of production and
progress, of that factor or element—and this factor
is the great man—which Karl Marx, with his doctrine
of labour as the sole creator of value, had eliminated.
Under one disguise or another the great or exceptional man, as distinct from the average labourer
whose productivity is measured by time, has been
put back in the place from which the theory of Marx
had ousted him; and the inventors, the men of enterprise, the organisers and capitalists of to-day—or, as
Mr. Sidney Webb calls them, "*the monopolists of
business ability*"—are given back to us in the guise of
officials of the bureaucratic State, armed by the State
with the industrial powers of slave-owners. It is
true that socialistic theorists still do their utmost to
hide from themselves and their followers the nature

of this change, by means of those curious arguments which find their chief exponent in Mr. Spencer, and which have rendered sociology thus far so useless as a practical science. But the change is but partly hidden, nevertheless, even from themselves.

Why, then, should they endeavour to hide it at all ? Why should they shrink from a perfectly frank avowal—an avowal which they are constantly com- pelled to make by implication — that the great man's power in wealth-production is what has been described, and that every increase in the wealth of civilised communities is due to him ? They shrink from making this avowal for one reason only. This reason is that their main practical object is to repre- sent the possessions of the great man, or of the few, as a treasure to which the few have no theoretical right, and which can be, and ought to be, divided amongst the many. They are therefore compelled, by the necessities of popular agitation, to obscure the part that the few have played in producing it, and to pretend, so far as possible, that it is produced by the undifferentiated many. If it were not for its promise to the many of some indefinite pecuniary gain, it may safely be said that socialism would have been never heard of ; and if this pecuniary promise were made good, the demands of the socialists, as a practical party, would be satisfied.

And now having considered this, let the reader look back at the claims that have, in our present argument, been advanced for the great man thus far. It will be seen that not a single claim has been advanced on his behalf to which, on practical grounds, any socialist could object. We have not assumed that out of all the wealth he produces he shall take a larger, or even so large a share, as the least efficient of his workmen. On the contrary, we have assumed that his contributions to the national wealth find their way into the pockets of those around him, and that for him nothing is left but the bare means of subsistence. It has indeed been shown that he must necessarily have the control of capital, and be free to use it in the way that he thinks best; but this is only because the control of capital affords the sole means by which, amongst free men, industrial discipline can be enforced and the productive genius of the few be communicated to the muscles of the many. For all that has been said thus far to the contrary, the great man himself may derive from his control of it no advantage whatsoever. We have assumed only that by his use of it he shall concentrate his exceptional faculties on the practical business of wealth-production with as much intensity and devotion as he would do if the whole of what he produced were to go into his own coffers. We have, in fact, been regarding the great man as being socially the servant of the ordinary men, though in technical matters he is their master.

So far, then, as our argument has up to this point proceeded, we have merely in our theory assigned to the great man functions which are implicitly assigned to him in the reasonings of the more recent socialists themselves, whilst in practice we have

Book III
Chapter 3

been made for the great man to which socialists need object;

for we have assumed that he keeps none of the exceptional wealth he makes, for himself,

assumed the realisation of the very conditions at
which socialism aims. For let us consider very
briefly what these conditions are. The more carefully the theoretical admissions and the practical
promises of the more recent socialists are examined,
the more clear does it become that the sole essential
change which socialism would introduce into the
existing economic *régime* would consist not in getting

but that he
works exactly
on the terms
the socialists
would dictate
to him.
rid of the great man, but in securing his activity on
totally new terms. The socialists aim, in fact, at
securing the best industrial masters and treating
them like the worst servants. This, as social
reformers, is their fundamental peculiarity. For
whilst they propose to secure an equal distribution
of products, they implicitly admit that the producers
may be divided into three classes—the men of exceptional ability who produce an exceptional amount
of wealth ; the mass of average men who produce
a normal amount ; and the idle, the refractory, and
the worthless, who produce less than the normal
amount ; and they propose accordingly to apportion
the products as follows. To the average man they
would give twice as much as he produces ; to the
idle and the worthless man they would give a hundred
times as much as he produces ; and to the great man,
on whose talents the fortunes of all the others depend,
they would give from a hundredth to a thousandth
part of what he produces.

Now, whatever the reader may think of this
economic programme, there is nothing in the present
work, thus far, to show that it is impossible ; and if

the object of socialists is to level social conditions,
to abolish all differences of rank, and to confiscate
all exceptional incomes, this book up to the present
,point might be accepted as a handbook of socialism.
For the reader will recollect that when it was said
that the great man's activity involved the existence
of motives which would lead him to develop his
faculties, and that without such motives these faculties
would be practically non-existent, the question of
what these motives were was for the time altogether waived, and we assumed the development
and the subsequent exercise of his abilities as
something that would take place no matter under
what conditions. The question, however, which
we then put on one side must now be taken up and
submitted to a careful examination. It being granted
that the activity of the great man is necessary, on
what conditions can his activity be secured ? Can it
be secured on the conditions that are proposed by
socialism, or on .any others that even remotely
resemble them ?

BOOK IV

CHAPTER I

THE DEPENDENCE OF EXCEPTIONAL ACTION ON THE
ATTAINABILITY OF EXCEPTIONAL REWARD, OR
THE NECESSARY CORRESPONDENCE BETWEEN THE
MOTIVES TO ACTION AND ITS RESULTS.

IN entering on the inquiry which now lies before us *Great men differ from ordinary men in degree only, not in kind,* it is necessary to recall to the reader, and to insist with renewed emphasis on a fact which has been ex- plained with the utmost fulness already. This is the fact that those exceptional efficiencies of the few on which the initiation, the progress, and the maintenance of civilisation depend, and which in a technical sense we have here described as *greatness,* do not consist of qualities which are unique in kind, or which are not possessed in some measure by the masses of ordinary men ; but that they are made up of ordinary faculties magnified or mixed together in unusual proportions. For although, as George Eliot observes in a striking passage, the faculties of all men are the same in kind, they manifest themselves in different men in such very different degrees that a faculty or feeling which in one man has the power and dimensions of a tiger, may never in another man outgrow those of

a weasel. *Greatness*, then, is simply the possession and exercise by such and such a person, in an exceptional degree, of some faculty or assortment of faculties, the rudiments of which are possessed by all. And the reason why it is necessary to insist on
this fact here is that, as a consequence of it, the use which the great man makes of his exceptional powers—or, in other words, their whole efficient existence—depends on certain causes which are relatively, though not absolutely, similar to those on which depends the use which the ordinary man makes of his.

Let us, then, consider the powers of the ordinary man first, and let us take as examples of them those powers or faculties which are most universally distributed amongst the human race—namely, the powers by which the rudest populations obtain enough⁚ food to live upon. Now such faculties, practically universal as they are, would be potential only, not actual, if it were not for two things. These are certain appetites or desires, having a physiological origin, on the one hand, and the external conditions on the other, which make the satisfaction of those appetites, or the fulfilment of those desires, a possibility. Thus if men could live without eating, and had no desire for food, those special faculties would
be dormant which are now exercised in agriculture ; and this means that for all practical purposes they would not exist at all. These faculties would also not exist at all, no matter what men's desire for food might be, if the whole of the earth's crust had

happened to be cast-iron, and if tillage were consequently impossible, and there were no seeds to sow.
In other words, the very commonest and very
simplest faculties which human beings possess have
a practical and a universal existence in those beings,
only because, in the first place, they minister to Thus the
exercise of the
universal wants, and because, in the second place, simplest facul-
the earth is so constituted as to supply the materials ties depends
on the want
on which these faculties can operate. Or, to put the of some certain
object, and on
matter in more general terms, the very commonest the possibility
of attaining it.
and simplest faculties are not practically self-existent,
except as mere barren potentialities ; and as practical
forces they exist only in the degree to which they
are evoked by external things and circumstances—
by some external object, such as food, which excites
and will satisfy desire, and by external circumstances
which make the object obtainable.

Now if this be true of those faculties of the com- If this is true
of the com-
monest kind, ministering to the needs which all men monest facul-
inevitably feel alike, and which they always must ties which aim
at supplying
feel so long as they remain alive, it is yet more necessaries,
much more is
obviously true of those higher and rarer faculties it true of rare
faculties, which
ministering to needs which are so far from being aim at producing super-
inevitable, that whole races have existed and do fluities.
exist without any conscious knowledge of them. The
great inventor, the great director of industry, will not
develop or use his exceptional latent faculties unless
by the use of them he can achieve some object which
he desires ; and this must be something which the
community has to give, or the possession of which it
will secure to him if it be something which he himself

produces. Columbus, for instance, as the records of his life show us, would never have braved the Atlantic if the society of his time, though in the end it rewarded him ill, had not rendered an enormous reward both in money and rank possible—a reward which he specifically bargained for in the event of his enterprise being successful. And similarly in the case of

great men in general, unless society is so constituted as to render some reward or other the natural or possible result of the exercise of certain exceptional faculties, and unless this reward shall be one which the great men shall think worth working for, their exceptional faculties will remain potential only. That is to say, their faculties will be practically non-existent, and the community will be as helpless as it would be if it had no great men at all.

Now here we have what is virtually a genuine social contract. It is not, indeed, such a contract as Rousseau dreamed of. It was never made deliberately at any period of history by two independent parties coming together for the purpose. It was the result of a gradual and quite unconscious process. Ordinary men, having experienced the advantages of being directed by great men, submitted instinctively to such conditions as the great men demanded, and instinctively offered them, or allowed them to retain possession of, such rewards as were necessary to stimulate them to further action. But these proceedings were a bargain, a social contract none the less, although they were not recognised as such ; and they constitute a bargain still—a bargain which is continually being renewed, and the terms of which reformers are continually trying to alter. Thus the socialists' proposal to take from the founder of a new industry all the wealth that his exceptional faculties have created, and pay him, as they propose to do, with the paper money of honour, is merely an attempt to make a new bargain with the great man, which shall secure his services on cheaper terms for the little men. Similarly, all encouragement offered to art and science by the State is a bargain offered to a number of unknown persons, who are presumed to be the possessors potentially of artistic and scientific faculties ; the State engaging to give them certain opportunities and rewards, if they on their part will make their potential faculties actual.

Now with regard to this bargain or contract which the community has not only made, but is always remaking and revising with its great men, we must observe that it is a bargain which, from the necessities of the case, is made by the community solely with individual great men who are living. It is not a bargain offered to the great men of the past, no matter how much of his greatness the living great man may owe to them. It is impossible to bargain with the dead, and therefore to the present question the claims of the dead are as irrelevant as the claims of protoplasm. The present question is how shall such and such living people be induced to develop certain superiorities which are latent in them, or to use to the best advantage superiorities which have been developed already. And

and this is a contract which is being constantly revised.

the answer depends on these men themselves. It depends on the characters which they personally possess, and not on the parents or ancestors from whom their characters have been derived. We can no more go behind the personality of the great man in bargaining with him, than we can go behind the personality of the dipsomaniac in attempting to cure him. We may excuse the failing of the latter as something which he has inherited from his ancestors ; we can cure it only as something for which he is himself responsible. If civilisation, therefore, depends

on the great man, no community can become or remain civilised which does not so arrange itself as to accord to its living great men such rewards as they themselves feel to be a sufficient inducement firstly to develop their faculties, and secondly to employ them to the utmost.

Here, then, we have a new and final verification of that truth which has already been established against the arguments of Mr. Spencer—namely, that the great man is a *vera causa* of progress, and that no explanation of progress has any practical value which does not base itself on an examination of the great man's character. And that such is the case will become yet more apparent when we take into consideration the following additional facts, which are quite distinct from any we have yet touched upon, and which practically have an equal, or perhaps even a superior, importance.

If the exceptional faculties of the great man were so far like the faculties possessed by all men,

that by looking at him we could tell that he was a
potential inventor, or organiser of industry, or philo- sopher, as easily as by looking at a common man we can tell that he can trundle a wheelbarrow, the entire force of the foregoing argument would be lost. The community would then know what each great man could do for it, and could force him to do it by flogging or starving him if he refused. The ordinary faculties—the faculties of manual labour—can be made to exert themselves precisely in this way. A large number of the great works of antiquity were due to labour successfully stimulated by the whip. But it is only a man's commonest faculties that can be called into action thus ; and they can be called into action thus only for this reason—that those who coerce him know that these faculties are possessed by him, and they also know the task which they wish to make him accomplish. But in the case of the great man both these conditions are wanting. It is impossible to tell that he possesses any excep- tional faculties till he himself chooses to show them ; and until circumstances supply him with some motive for exercising them, he will probably be hardly aware that he possesses such faculties himself. Moreover, even if he gives the world some reason to suspect their existence, the world will still not know what he can do with them, and will consequently not be able to impose on him any task until he himself chooses to show of what he is capable. Any farmer by looking at Burns could have told that he had the makings of a ploughman in him, and have forced

him, under certain circumstances, to do so much
ploughing daily; but no one could have told that
he was a poet if he had not of his own free will
revealed the fact to the public; and even when the
public were aware of it, no one could have forced
him to compose *The Cotter's Saturday Night.* A
press-gang could have turned Columbus into a
common sailor, but not all the sovereigns of Europe
could have forced him to discover a new hemisphere.
On the contrary, it was he who had to force sovereigns into the reluctant belief that possibly there
was a new hemisphere to discover.   The great man,
therefore, is lord of his exceptional faculties in a way
in which the common man is not lord of his common
faculties.   The existence of the latter faculties cannot be concealed; the kind of work that can be
accomplished by them is known to everybody; and
therefore the community by the exercise of mere
force can command the average man, and make him
work like an animal.   But over the exceptional
faculties of the great man it has no command whatever, except what the great man gives it; for it
neither knows that the faculties exist, nor what things
the faculties can do, until the great man elects to
reveal the secret.   He cannot be made to reveal
it, he can only be induced to do so; and he can be
induced to do so only by a community which offers
to exceptional faculties some assured and exceptional
reward, just as a reward is offered for evidence
against an unknown murderer.   Moreover, just as
in the latter case it very often happens that the re-

They cannot,
therefore, be
coerced from
without, like
ordinary
workers.

They must be
induced to
work by a
reward

ward originally offered has to be raised several times before a sum is reached which will induce the witness to come forward, so must any community, as the condition of becoming civilised, raise the rewards of which they themselves feel greatness to such a figure that the possessors of to be sufficient. latent superiorities will be induced to develop and use them. And hence the great man not only causes progress by what he does, but he influences also the entire structure of society, by his character, which regulates the terms on which he will consent to do it.

This is the point at which the science of sociology Hence the great man's primarily comes in contact with the practical prob- character and requirements lems of to-day. That all progress is due to the impress them- efforts of the superior minority is a truth which, selves on the structure of taken by itself, and apart from other truths allied to it, society. we can merely recognise and assent to. We can do nothing to alter it ; nor will the fact of our recognis- ing it, if taken by itself, tend to alter or guide our conduct. We are not even able to settle the number of males and females which shall be produced in each family. Still less can we settle or increase the number of individuals who shall bring into the world with them talents more than ordinary. But though no community can do anything to settle or alter the percentage of potential greatness that will be born into it from generation to generation, it can settle or alter the social conditions and rewards by means of which this potential greatness shall be developed and enabled to use itself ; and a very large part, though not the whole, of political wisdom

will thus consist in arranging these conditions and
rewards, so that from each potentially great man,
whatever degree or kind of potentiality may be his,
the community may elicit the highest and most far-
reaching efforts of which he is capable. It will, of
course, be to the interest of the community to secure
this result by offering the great man the smallest
and least costly reward, the desire of which will
induce him to develop and exert himself to the
utmost; but the ultimate fixer of the great man's
price—let it once again be said—is not the community, but the great man himself.

It is this sociological and psychological truth
that even the clearest-headed amongst the socialists
are continually forgetting. They perceive it at one
moment, at the next moment they entirely forget
it, and solemnly proceed to build up their visionary
polity on foundations which their own arguments
had previously condemned. A curious example of
this "*inability*," as Mr. Spencer calls it, "*to comprehend assembled propositions in their totality*" is
to be found in a remarkable passage by Mr. Sidney
Webb. Having observed that "*socialists would
nationalise both rent and interest by the State becoming the sole landowner and capitalist*," he goes on to
acknowledge that great fundamental fact which it is
the main object of the present work to elucidate.
"*Such an arrangement, however,*" he says, "*would
leave untouched the third monopoly—the largest of
them all—the monopoly of business ability.*" In
these last words he appears to be like a Daniel

come to judgment. He recognises in the fact that the few have a natural monopoly of faculties, the exercise of which is required for the progressive well-being of all, a genuine and a formidable difficulty in the way of the realisation of socialism; but now comes the passage for the sake of which these others have been quoted. Great as this difficulty is, he tells us, "*the more recent socialists*" have devised a way for getting over it. And what does the reader think this way is? It has at all events the merit of being very simple. "*The more recent socialists*," says Mr. Webb, "*attack this third monopoly also by allotting to every worker an equal wage, whatever may be the nature of his work.*"

It has been thought worth while to quote Mr. Sidney Webb because he is an exceptionally favourable specimen of the modern socialistic theoriser. It is therefore interesting to notice the hiatus that here yawns in his argument. The entire question which is really at issue is begged by him. His allies, he tells us, though they cannot destroy the monopoly which the few possess of exceptional business powers, will destroy the effects of this monopoly by taking away from the few nearly all the wealth that their exceptional powers produce. It never seems to occur to him to ask whether, under these circumstances, the few would develop or exercise their exceptional powers at all. And yet the whole problem for him, as a socialist, lies here, and lies nowhere else. For from the very fact that these powers are admittedly a monopoly of the few, it is

and they propose to equalise matters by not offering great men any exceptional reward.

They forget to ask whether, under these circumstances, great men would exercise or reveal their exceptional powers at all.

evident that their existence cannot be assumed in anybody unless he exerts himself to give some sign of their presence. External authority, therefore, can compel nobody to employ them who does not put himself at the mercy of the authorities by letting them know he has them; and thus "*the more recent socialists,*" in attacking "*the third and greatest monopoly,*" are really themselves at the mercy of the very monopolists whom they propose to attack. It is true that if a socialistic revolution could be brought about suddenly, existing great men known to have certain talents, which had been already developed and exercised under conditions which the revolution destroyed, might be seized on by the State, in its capacity of universal employer, and forced to continue something of their former voluntary activity by threats of torture or some similar method of coercion. But even granting this to be possible, it would only solve the problem for a moment; for as these men died— and some of them would be dying daily—new talent

Exceptional rewards are essential to exceptional action. would be wanted to take the place of the old; and though the State might coerce such talent as was already developed, it could not by coercion secure the services of the new, because threats of coercion would never tempt new talent to discover itself, but would, on the contrary, drive it yet deeper beneath the surface.

Exceptional potentialities can be called out and realised only by a kind of action which is the very antithesis of coercion, and which is analogous to that of sunshine on buds, or flowers or fruits

— namely, the penetrating, the warming, the stimulating action of the hope of certain personal advantages on the mind of the exceptional man, which advantages he will not only covet as advantageous, but will recognise as the natural result of the exercise of his exceptional faculties, and as a result attainable by the exercise of these faculties only. What these personal advantages are, the desire of which, coupled with their attainability, is necessary to stimulate men who have more than ordinary potentialities, to do greater things by developing them than are done by ordinary men, must be determined by reference to the actual facts of life, the records of which are ample, and the details of which, though numerous, can by careful analysis be easily reduced to order.

Book IV
Chapter 1

We must inquire what the required exceptional rewards are.

CHAPTER II

THE MOTIVES OF THE EXCEPTIONAL WEALTH-PRODUCER

Socialists, though often forgetting the necessity of exceptional motive, often remember it, IN spite of their frequent forgetfulness of the fact just insisted on, that the development and exercise of exceptional faculties can be secured only through the influence of some exceptional motive, this is not a fact which socialists theoretically deny. On the contrary, often as they forget it, with curious consequences to their reasoning, yet just as often, when they happen to be directly confronted with it, they are loud in declaring that they recognise it quite as clearly as their opponents ; and a considerable portion of their more modern writings consists of a setting forth of the various exceptional rewards which will, according to them, in the socialistic State, *and endeavour to show that socialistic society would have sufficient rewards to offer to its great men,* elicit from exceptional men the exercise of their utmost powers. Moreover, the rewards on which the socialists principally insist are rewards, the desire of which is admitted by all parties to be an actual force in society as at present constituted, and in fact to have been, ever since the dawn of history, the motive to which much activity of the highest kind has been due. These rewards have been defined in a recent *Handbook of Socialism* as the pleasure of "*excelling*," "*the joy in creative work*," the satis- faction which work for others brings to "*the instincts of benevolence*," and, lastly, "*social approval*," or the homage which is called "*honour*."

If the socialists, however, confined themselves to maintaining that the desire of such rewards as these constitutes a sufficient motive to exceptional activity of certain kinds, they would not only be asserting what nobody else would deny, but they would be putting forward nothing which, as socialists, it is their interest to assert. The ultimate proposition which, as socialists, they aim at establishing is not that certain kinds of exceptional men do certain kinds of exceptional things, in obedience to the motives in question ; but that because some excep- tional men, endowed with certain temperaments, are motived by them to activities of certain specific kinds, other exceptional men will be motived by them with equal certainty to other activities of a kind totally different—and more especially to the activities which result in the production of wealth.

Here is the fundamental point on which the socialists join issues with their opponents. Their opponents, they say, assume that the sole reward or advantage, the desire of which will stimulate the monopolists of "business ability" to exert that ability in the production and augmentation of wealth, is a share of wealth for themselves pro- portionate to the amount produced by them—an amount which will separate their lot from that of the majority of their fellows. Now if this should be really the case, as the socialists are coming to perceive, the fact would be fatal to the entire ideal of socialism. They are consequently now directing the best of their ingenuity to showing that the

Is the enjoy-
ment of excep-
tional wealth
superfluous
as a motive to
producing it?
desire of possessing exceptional wealth is altogether superfluous as a motive for producing it, and that the great producers of it, when all chance of possessing it is taken from them, will find in the pleasures of the strain which the productive process necessitates—especially if these are supplemented by the inexpensive thanks of the community—a more powerful inducement to exertion than is the prospect of the largest fortune.

If it is so, it is
for the social-
ists to prove
that it is so ;
Now in endeavouring to make this peculiar position good, it is evident that the burden of proof lies with the socialists themselves ; for although the doctrine that all exceptional exertions in wealth-production are motived solely by an avidity for exceptional wealth as such—and this is the doctrine which the socialists set themselves to controvert—is a very imperfect rendering of what their opponents actually maintain, it embodies an assertion which the socialists themselves declare to have been true of all exceptional exertion in wealth-production hitherto. No one declares this more passionately and more persistently than they. For what, as political agitators, has been their chief moral indictment against the typical great men of industry — the organisers of labour, the introducers of new

machinery, the pioneers of commerce ? Their chief Book IV
moral indictment has been this : that these men, Chapter 2
instead of labouring for their fellows, or for the for they them-
sake of any of those rewards which the socialists selves admit that it has not
declare to be so satisfying, have been motived been so in the past, and is not
solely by the passion of selfish "greed." Its hideous actually so now.
influence, they say, is as old as civilisation itself, and
the "*monopolists of business ability*" in Tyre and
Sidon were as much its creatures as are their
modern representatives in Chicago. And this assertion, unlike many made by the socialists, has the
merit of being, so far as it goes, true. Greed, of
course, is a word which, in addition to its direct
meaning, carries with it an accretion of moral
insult ; but putting aside this, it means in the present
connection merely a desire on the part of the great
wealth-producer to enjoy an amount of wealth proportionate to the amount produced by him : and
from the dawn of civilisation up to the present time
all great wealth-producers, whether merchants, manu-
facturers, or inventors, have had the desire of enjoy-
ing such wealth as their motive. The desire has
been connected with the activity just as universally
and closely as the desire of water is connected with
the act of drinking it, or the desire of winning a
woman with the act of making love to her. If the
socialists, then, would persuade us that a motive so
universal as this can be now superseded by others of
an entirely opposite character, they can do so only
by adducing the clearest evidence that, on the one
hand, this motive itself is losing its old power, and

Book IV
Chapter 2

Are there any
signs, then,
that the desire
for exceptional
wealth is
beginning to
lose its power?

We shall find
that the
socialists
themselves
maintain just
the contrary ;

that other motives, on the other hand, are actually acquiring and exercising it.

Let us first, then, consider the passion of greed itself, and ask whether there is anything in its connection with wealth-production hitherto which may lead us to think that in spite of its universality in the past, it is merely a transitory propensity from which exceptional men will free themselves, instead of being a propensity rooted in the very constitution of human nature.

And here again the socialists will be amongst our most important witnesses; for just as they, of all writers and thinkers, have done most to call attention to the fact that up to the present time greed has been the main motive by which the exceptional wealth-producers have been actuated, so they, of all writers and thinkers, have done most to call attention to another fact as well, which shows the motive in question to be as permanent as it is universal. For that very desire of the producer to possess what he himself produces, which, when found in the exceptional man, they denounce as greed, and which they tell us that the exceptional man will get rid of in the course of a year or two, is the very desire which, as existing in the common man, they have assumed to be the foundation of his whole industrial character; and to it have all their most fervid and powerful appeals been made. The socialists, in their attempts to excite the masses against the existing order, have relied less on rhetorical declarations that the labouring man gets very little, than on the quasi-scientific assertion that he gets less than he produces, and that consequently the wealth of his employers is merely his own wealth stolen from him. "*All wealth is due to labour ; therefore to the labourer all wealth is due*" has formed from the first, and still forms the text from which the socialists always preach when addressing the labouring classes ; and the use of this text as the watchword of popular agitation is obviously an admission that, as a producing agent, man is motived so exclusively by the desire to possess what he produces, or else its fair equivalent, that he naturally resents the idea of producing anything merely in order that others may take it away from him. Indeed, this doctrine that the desire for the product, and the producer's sense that he has a right to it, form the only motive for production possible for a free man, formed the unquestioned basis of the entire socialistic psychology so long as the theory of Marx was held by the socialists to be unassailable, according to which wealth was the product of average labour, and the common or average labourer was the sole true producer. It was only as time went on, and the socialists were slowly compelled to recognise the few to be producers of wealth just as truly as the many, that the socialists began their attempts to get rid of the doctrine which a very little while ago they regarded as axiomatic—the doctrine that each producer has a right to his own products, and that his hope of possessing it is his principal motive for its produc-

for they appeal to the desire of each producer to possess all he produces as the most universal and permanent desire in man ;

and never questioned this so long as they believed that the sole producer was the labourer.

19

tion.   In making these attempts, however, they have, with a judicious eclecticism, been content to apply them to the exceptional man only; and the common man and his motives they leave undisturbed, except when they venture on the doctrine that the common man's motive for production will in 'the future be the desire of possessing, not only all that he produces, but all that he produces and a great deal else besides.

If, then, it is unlikely that this desire to possess the product will cease to be operative as the motive to production amongst the masses, that it will cease to be operative amongst the few is more unlikely still ; for the man who is possessed of average powers only, cannot hope to produce more than the average man requires, and his object in producing tends to represent itself to his mind in terms of the comfort which he hopes to experience, rather than in terms of the value of products which he hopes to possess.   But the exceptional man, whose peculiarity as a producer is this, that he produces not only as much as the average man requires, but an indefinite amount in addition to it, is constantly balancing his products not with his immediate wants, but with the amount of intellectual effort which he has expended in the process of production.   Indeed, the more closely we consider the matter, the more strongly we shall be convinced that the desire of possessing wealth proportionate to the amount produced by them becomes as a motive to production stronger in men, not weaker, in exact proportion as their productive powers are great, and the amount produced by them appeals to their intellects rather than to their necessities.

So far, then, as a study of this motive itself can inform us, the socialistic idea that it will ever cease to be paramount has no foundation whatever, and is con-tradicted even by the socialists themselves. The only fact connected with this motive directly which wears so much as a semblance of serious evidence in their favour is the fact often dwelt on by emotional writers like Mr. Kidd, that many men who have made enor-mous fortunes have given away a large part of them for what he calls "altruistic" purposes; and writers of the kind in question take this fact for evidence that the desire of possessing great wealth is ceasing to be the motive for producing it. But those who allow themselves to argue thus, show a curious carelessness in their examination of human action; for the fact referred to, so far as it proves anything, negatives rather than supports the conclusion they seek to draw from it. It is perfectly true that many men of great industrial ability have produced large fortunes and given them away afterwards. But in order to give, a man must first possess; and it is in the act of giving magnificently for some specified purpose that many men most fully realise the power with which wealth endows them. Thus the fact that many men will produce in order that they may have the delight of giving is no more a proof that they would produce under the *régime* of social-

for even if he gives away what he pro-duces, he desires to possess it first.

ism, which would aim at depriving them of anything that they might possibly give, than the fact that a man would with pleasure give five shillings to a beggar is a proof that he would be equally pleased if the beggar were to pick his pocket.   Even the men who produce wealth—and no doubt there are such—without any conscious sense that they produce it because of their desire to possess it, would show that such was their motive by their instinctive and indignant refusal to go on producing it, if they knew that it would be forcibly taken from them.

There is no sign, therefore, that the desire for exceptional wealth is losing force as a motive.   And now, since we have seen that "greed" as a motive to wealth-production shows no internal tendency to lose its old efficiency, let us turn to those other motives which the socialists tell us are to supersede it, and ask whether there is anything in their known operations hitherto which indicates that in the domain of wealth-production they will acquire an efficiency similar to it.   This is not an inquiry which is very difficult to pursue, for the motives in question are of a very familiar kind, and the kinds of activity which they have produced hitherto are notorious.

Are, then, other desires acquiring new force as motives to wealth-production?   What these motives are has been sufficiently shown already in language borrowed from the socialistic writers themselves—the pleasure of "*excelling*," the "*joy in creative work*," the pleasure of doing good to others, and, lastly, the enjoyment of the approbation of others, or of the yet more flattering tribute commonly called "*honour.*"   Now these motives, it will be seen, are of two distinct kinds, the first three

being based exclusively on some pleasurable condition of mind, which is independent of anybody
except the individual who actually experiences it ; Are the joys of
excelling, or of
the two last being based on a pleasurable condition benefiting
others, or of
of mind, which is directly dependent on the actions being honoured
by others
or the attitude of other people. We may therefore doing so?
reduce these motives to two—namely, self-realisation, in the first place, and recognition by others, in
the second. This classification will be not only
shorter, but more comprehensive than the other ;
for self-realisation will include not only the joys of
self-improvement and artistic creation, but those of
the pursuit of truth and the performance of religious
duty, and will distinguish the pleasure of doing
good to others from the pleasure of being thanked
or praised for it.

And now let us consider what those kinds of The desire of
these joys is a
exceptional activity are, in the production of which motive to
certain kinds
one or other of these motives, or both of them, of exceptional
have played, hitherto, any considerable part. We conduct.
shall find them to be as follows : heroic conduct
in battle, or in the face of any exceptional danger ;
artistic creation ; the pursuit of speculative truth ;
what theologians call works of mercy ; and, lastly,
the propagation of religion. This list, if understood
in its full sense, is exhaustive.

Now of these five kinds of action we may dismiss It is a motive
to benevolent
the last from our consideration, not because it has action and
religious work;
not a most important influence on civilisation, but
because it has no direct connection with any of the
processes of wealth-production, except in so far as it

294 ARISTOCRACY AND EVOLUTION

Book IV
Chapter 2 tends to divert men's attention from them. And with regard to the works of mercy something similar must be said also; for though they undoubtedly have a close connection with wealth, they do not aid at its production, still less at its increase, but merely at the distribution of portions of it, which have been produced already, amongst persons whom it would otherwise not reach. The love for others, for example, by which works of mercy are motived, may prompt a man to send London children for a holiday into the country by train, but it would never have prompted him to invent the locomotive engine. It may prompt him to secure for a youth an education in but neither of these is the same thing as wealth-production. modern science, but it would never have prompted him to write the treatises of Professor Huxley. All activity of this kind, then, whatever form it may take, is, in a sociological sense, essentially parasitic. It implies the previous exercise of another set of faculties totally distinct from those directly implied in itself, and, together with other faculties, other motives belonging to them. It has, then, with the actual process of wealth-production as little to do as has religious propagandism itself; and, like religious propagandism, we may dismiss it from our consideration here. The only forms of activity with which we are called on to deal with here will thus be artistic creation, the pursuit of speculative truth, and military or quasi-military feats of heroism.

It is a motive to artistic production, certainly, As to artistic creation, it is, no doubt, perfectly true, as is proved by the efforts of countless devoted amateurs, that men with artistic powers will

often do their utmost to develop them, merely for
the sake of the pleasure which the exercise of these powers brings with it ; whilst literature is even more obviously than painting cultivated by men who devote themselves to it solely as a means of self-expression. Indeed, it might reasonably be contended that finer books and paintings would be produced if it were impossible for painters and writers to make money by producing them, than are now produced with a view to captivating the public purchaser.

So, too, the pursuit of scientific and philosophic truth—arduous though it is—is generally under-taken by men whose principal motive is the pleasure their work brings them.

> A watcher of the skies,
> When some new planet swims into his ken,

may well be supposed to find in that thrilling moment a reward sufficient to compensate him for all his pains in arriving at it ; and most branches of science would yield us similar illustrations. Indeed, the career characteristic of scientists and philoso-phers generally is a conclusive proof that the principal motive of their activity is not the desire of any extrinsic reward, the amount of which they will balance against the amount or the quality of their efforts, but a passion for truth as truth, which they indulge in for its own sake only.

Now granting all this, what will its bearing be on the question of whether the pleasures of pure self-realisation will suffice to stimulate those exceptional faculties whose function it is to maintain and increase the production of wealth ? With regard to artistic creation, we are certainly bound to admit that great works of art are wealth of a highly important kind, and when a good picture is produced, as it often is, solely in obedience to the painter's artistic impulse, we have a genuine example of wealth produced in obedience to that kind of motive whose efficiency the socialists desire to establish.   Further, with regard to the pursuit of truth, as Mill points out in a passage that has been already quoted, progress in speculative knowledge is the basis of all other progress, and notably of progress in the arts and processes of wealth-production.   It must, accordingly, be admitted that in a certain sense all progress in wealth-production has for its basis a kind of disinterested activity with which the desire of possessing wealth has nothing at all to do.   And yet in spite of this, neither the case of the artist nor of the philosopher warrants the inference that the motives which are sufficient for them will ever have a similar effect on the faculties of the great wealth-producers.   The evidence, in fact, as soon as we have fully examined it, will be found to point in a direction precisely opposite.

and works of
art are wealth ;
and scientific
discovery is the
basis of
industrial pro-
gress ;

but great
works of art
form but a
small part of
wealth ;

For, to begin with the case of the artist, it must be remembered, in the first place, that works of art, such as pictures painted by the artist's hand, form a very small, though an important part of wealth, and that they are hardly wealth at all from the

point of view of the many, unless they are reproduced and multiplied by adequate mechanical
processes. Now, though it is quite conceivable
that a painter might paint a Madonna solely
because the realisation of his own ideas delighted
him, it is hardly to be expected that other men will
rack their brains to devise blocks, presses, and pre-
parations by which copies of it may be made and
multiplied, solely for the pleasure of reproducing
ideas which are not their own. It must further be
added that delight in creation for its own sake can
be attributed as a sufficient motive to the highest
class of artists only. As for the men whose artistic and artistic
powers are true, but qualify them only for decorative effort other
than the
not for creative work—the men, for example, who highest is
motived by the
design beautiful stuffs and furniture—though the desire of
pecuniary
exercise of their power may be doubtless itself a reward;
pleasure to them, they are certainly as a class not
given to exercising them without the expectation of
some proportionate pecuniary reward. Indeed, in
exact proportion as artistic creation assimilates itself
to the processes by which wealth in general is produced, the mere pleasure of the work itself ceases to
be a sufficient motive for it.

Next, with regard to the pursuit of speculative whilst scientific
discoveries,
knowledge, though this, and more especially pure though made
generally from
scientific discovery, may form the basis of all pro- the desire for
truth, are
ductive effort, it is very far from being a form of applied to
productive effort itself. It has, on the contrary, no wealth-production because
necessary connection with it. It does not even the men who
apply them
belong to the region in which such effort operates. desire wealth.

Scientific truths, as apprehended by the mere seeker after speculative knowledge, are like powerful spirits secluded in some distant star; and, for any effect which they have on the processes of economic production, they might just as well have never been discovered at all. Before they can be applied to practical purposes they have to be mastered and digested by a new class of men altogether, who value them not for themselves, but solely for the use they can be put to. Thus, in order that speculative truths may be connected with productive effort, they must pass out of the hands of the men who first discovered them, and be made over to men whose motive in acquiring them will emphatically not be desire of the mere pleasure of intellectual acquisition, but the desire of some marketable products with a calculable pecuniary value, in the production· of which a knowledge of the truths in question will help them. Thus speculative activity, just like artistic creation, in exact proportion as it connects itself with the ordinary processes of wealth-production, ceases to find its motive in the desire of self-realisation, and claims to be rewarded by the possession of the objective results produced by it.

What, however, of the fact that the desire of honour makes the soldier work harder than any labourer?
And now let us turn from the motives which consist in the desire of self-realisation to those which consist in the desire of the approbation or the homage of others. This desire, which exercises a great influence on the artist, and often also on the seeker after speculative truth, concurrently with the

desire of pure self-realisation, exhibits its force most signally when it is the motive of military heroism; and the readiness with which a soldier will risk his life for honour — honour which brings with it nothing besides itself, excepting perhaps a medal and a scrap of ribbon—has been said by socialistic writers to afford a conclusive proof that any practical work, no matter how laborious, and more particularly the work of the great wealth-producer, will be willingly undertaken for the sake of the same reward. *"The soldier's subsistence is certain,"* writes a well-known contemporary enthusiast. *"It does not depend upon his exertions. At once he becomes susceptible to appeals to his patriotism. He will dare anything for glory, and value a bit of bronze which is 'the reward of valour' far more than a hundred times its weight in gold."* The implication, of course, is that what men will do in war they will do in peaceful industry; and the writer adds, in order to point this moral, *"yet many of the private soldiers come from the worst of the population."* This passage is quoted with rapture by another socialistic theorist, who exclaims, *"Let those especially notice this last point who fancy we must wait till men are angels before socialism be practical."* And even so well-trained a thinker as Mr. Frederic Harrison has argued, from the readiness with which men die in battle for their country, that they will be equally willing to deny themselves or suffer martyrdom for universal humanity.

Why, the socialists ask, should not the same desire make the great wealth-producer work?

Mr. Frederic Harrison has urged a similar argument.

To all these ideas and arguments there is one

Book IV
Chapter 2

The answer to
this is that the
work of the
soldier is
exceptional; answer to be made. They are all founded on a failure to perceive the fact that military activity is in many respects a thing apart, and depends on psychological, and indeed on physiological processes which have no counterpart in the domain of ordinary effort. That such is the case can be seen very easily by following out the train of argument suggested by Mr. Harrison. Mr. Harrison sees that in ordinary life a man will not deliberately run the risk of being killed except for the sake of a cause or person to which or whom he is profoundly and indescribably attached. Indeed his attachment is presumably in proportion to the risk he is prepared to run. And such being the case in the field of ordinary life, Mr. Harrison assumes it must be the case on the field of battle also, and that the soldier's willingness to risk death in fighting for a cause or country proves that this cause or country is inexpressibly dear to him. And in certain cases—when a country is in desperate straits, and everything hangs on the issue of a single battle—this inference would be doubtless just; but that it is not so generally is shown by the notorious fact that some of the bravest and most reckless soldiers ever known to history have been mercenaries who would fight as willingly for one country as for another. Thus until Mr. Harrison can show us that men in ordinary life will wear themselves out for either of two opposed objects indifferently, or that they will risk death as willingly for a plain woman as for a pretty one, it is obvious that men's willing-ness to risk death in war implies no corresponding

willingness to risk it cutting trousers, and is for
certain reasons a phenomenon standing by itself.

That this is so is shown even more strikingly by the fact to which the two other writers just quoted point with so much complacency. This fact is the soldier's undoubted willingness to pursue his calling for pay which seems strikingly incommensurate with his risks. His conduct in this respect is, no doubt, remarkable, especially when compared with that of men in the domain of peaceful industry. When any industrial occupation is in question a workman will expect special wages if it is one which presents a likelihood of his often hurting his thumb ; but soldiers will risk the probability of being tortured and blown to pieces for wages which would hardly induce a peasant to hoe a turnip-field. This is no indication of any abnormal poverty amongst the classes from whom the army is mainly recruited, for the same phenomenon is constantly observable amongst men who are not under the necessity of working for their living at all. Amongst such men are numbers who in time of actual war will eagerly give up a life of leisure and luxury for the certainty of hardship and the probability of death—men who for the sake of anything else but fighting would hardly, without a struggle, run the risk of a bad dinner. But what these facts really suggest to us is not the insane conclusion that because soldiers act differently from other men, other men may be counted on to act like soldiers. On the contrary, what they suggest is the question why men will do as soldiers what no one will do in any other capacity, and what soldiers themselves will cease to do as soon as they become commissionaires.

For this peculiarity in the soldier's conduct there are three separate reasons. One is the strictness of military discipline, which socialistic reformers would hardly find popular if they tried to introduce it into factories and contractors' yards. A second is the peculiar character of the circumstances in which the soldier is placed when his courage is most severely taxed—circumstances which render the attempt to evade peril almost as difficult, and often more perilous than facing it, and which in ordinary life would be intolerable if they did not happen to be impossible. But the most important reason is this —and the others without it would be non-existent —that the instinct of fighting is inherent in the very nature of the dominant races, and it will always prompt numbers to do for the smallest reward what they could hardly, in its absence, be induced to do for the largest. This immemorial instinct has been wrought into our blood and nerves by the innumerable thousands of years that have made us what we are ; and all the battles of their fathers are pulsing in men's veins to-day. These instincts, no doubt, are more controlled than formerly, and not so frequently roused ; but they are still there. They are ready to quicken at the mere sound of military music ; and the sight of a regiment marching draws cheers from the most democratic crowd. Here is the

reason why the soldier, though he submits himself
to the most direct coercion, never considers himself, and never is considered a slave ; and military activity will always be a thing apart, and for *in a way in which the* purposes of argument will never be comparable to *industrial* industrial, till human nature undergoes so radical *instinct is not.* a change that men will as eagerly risk being killed by unfenced machinery in a cotton-mill as they will being killed by a bullet or a bayonet on the field of battle. Here again the facts for which the socialists reason are indubitable ; but the inference which the socialists draw from them is altogether illusory.

It remains, however, to add that the desire of *And even in war those who* mere honour — of honour unaccompanied by any *make the pro-* extrinsic advantages — has an efficiency which is *longed intel-lectual efforts* strictly limited in the domain even of military *required, ask for other* activity itself. It may move men, in the act of *rewards besides mere honour.* fighting, to the highest and most heroic actions ; ~ ~ but history shows us that it has not been found sufficient to elicit the sustained intellectual efforts of the General, bent on achieving some great and monumental conquest — efforts in which all the excitement of the actual fighter is wanting, and in which the coolest calculation plays as large a part as courage. The Cæsars and Napoleons of the world have certainly not, as a rule, been content, when they have crushed their enemies and augmented the magnificence of their country, with the gift of a medal or two, and the privilege of ending their days in the modest uniform of commissionaires opening shop doors.   If, then, the mere honour of being a great conqueror is insufficient to stimulate the

Still more will
the great
wealth-pro-
ducers do so.
activities by which great conquests are achieved, a man is hardly likely to consecrate his entire faculties to wealth-production merely that he may enjoy the honour of being known as the proud producer of so many miles of calico, or millions of pots of jam.

There is there-
fore nothing
to show that
these other
motives will
supersede the
desire of
wealth.
There is, therefore, in the present operations of those motives, for which the socialists attempt to claim a universal efficiency, as little to suggest that as motives to exceptional wealth-production they will ever supersede the desire of exceptional possession, as there is in the present operations of the desire of exceptional wealth-possession to show that it is losing its power, or is at all likely to be superseded. The final demonstration of this truth, however, yet remains to be given.

What they
really do, and
what socialists
fail to see, is to
mix with the
desire for
wealth, and
add to its
efficiency ;
The socialists, in dealing with this question of motive, have been led into the curious blunders which have just now been exposed by their singularly childish conception of what men's actual motives are.   They divide motives into various well-known classes, and, so far as it goes, their procedure is here correct.   Their error is that they conceive of man as a being on whom these motives, as a rule, act separately ; whereas in reality the very reverse is the case.   Acts which are due to any single motive are not the rule, but the exception.   For instance, even though artistic creation and the pursuit of truth are motived in the case of many men by the pleasure which the work brings them, some of the greatest

artists and thinkers, with whom this motive was certainly powerful, have been motived by the desire of pecuniary reward also. It is enough to mention the names of Bacon and Shakespeare, Rubens, Turner, and Scott. And with the desire of honour the desire of pecuniary reward is found to mix itself yet more often and readily than it does with the mere passion for artistic or for speculative work itself. The psychological fact, however, which we must here notice is this—that the pecuniary reward, though it seems theoretically to be in contrast to any genuine desire for other men's approbation, or for the pleasure brought to the worker by the work itself, instead of destroying the force of those other motives, increases it, just as the admixture of a certain amount of alloy makes gold and silver more valuable for artistic purposes. And now, having observed this, let us turn back to the consideration of the desire of pecuniary reward as the principal motive of wealth-production, and endeavour to make our analysis of it more complete.

Book IV
Chapter 2

as the desire of wealth has mixed with other desires in men like Bacon, Rubens, etc.

As the reader will recollect, the doctrine that all exceptional exertions in wealth-production are motived solely by the desire of exceptional wealth as such, although it is the doctrine imputed by the socialists to their opponents, has been said already to be a very imperfect rendering of any doctrine as to the subject which their opponents would actually maintain ; and the reason why it is imperfect is simply that wealth as such is not the object for which wealth is really sought by most of those men whom the

For in saying that the desire of wealth is essential as a motive to wealth-produc- tion we do not mean the desire of wealth for its own sake, desire of it most powerfully influences. For wealth as such, in the ordinary sense of the phrase, is wealth regarded as a means of personal self-indulgence. It stands for the finest wines, the richest food, the softest beds, the most luxurious furniture—for everything that can caress the senses and enervate the mind and body. And no doubt its power of securing all these things to its possessors is one of the qualities which render it an object of desire. But it is only one; and though it is the most obvious of them, it is not the chief. The subordinate place which it occupies is conclusively shown by the fact that a very few thousands a year would suffice to provide a man with every pleasure or luxury that his own senses could appreciate ; and yet men are often more eager, after these few thousands have been secured by them, to pass this point of opulence than they ever were in reaching it. Many men, moreover, who have surrounded themselves with pomp and splendour are indifferent to the gratification of their own senses altogether. Though their luncheon tables may groan under every imaginable delicacy, they will themselves eat a slice or two of cold ham, no better or worse than would have been secured them for a shilling in a cheap restaurant. Their own beds will be no softer than those of prosperous clerks ; and, surrounded by cushioned sofas, they will sit upon straight-backed chairs.

The principal reasons for which wealth is sought are not pleasures of the senses, but pleasures of the mind and the imagination ; and of these pleasures there are three principal kinds. One of them is the pleasure of power, which in their analysis of human motives the socialists conveniently overlook; and the two others happen to be the very pleasures by the desire of which the socialists themselves declare the exceptional wealth-producers are to be principally marked in the future—namely, the pleasures of self-realisation and the pleasures of social honour. Wealth is coveted by all really great wealth-producers, not in preference to these, but as a means to all or one of them. To many of our great wealth-producers, with their strong practical faculties, wealth would be nothing if it brought to them no accession of influence; to many it would be nothing if it did not bring them the means of indulging their tastes, as distinct from their physical appetites; to nearly all it would be nothing if they did not, or if they did not hope it would, secure for them the approbation and the respectful homage of others.

The only alternatives, then, which we have before us are as follows :—If the great wealth-producer is a man of such coarse fibre that none of those desires just mentioned are really his—neither the desire of power, nor the desire of social honour, nor the desire for that larger development of taste and moral activities which is rendered possible by the possession of exceptional wealth—then it is obvious that the sole motive left to him will be the gross or unreasoning desire for the possession of wealth as such; and we are brought back to the original

Book IV
Chapter 2

It is desired mainly as a means to power, and to those very pleasures which socialists offer instead of it.

The great
wealth-pro-
ducers suscep-
tible to the
motives on
which the
socialists
dwell will
desire excep-
tional wealth
all the more
because of
them.

proposition which the socialists set themselves to annihilate. But if, on the other hand, the great wealth-producer is really capable of those higher desires which the socialists assure us will shortly become so strong in him, the desire of exceptional wealth, instead of being superseded by these, will be stronger beyond calculation than it ever could be without them.

And it is, as a rule, the latter of these two suppositions which practically represents the truth. Exceptional wealth is desired by the men who produce it not for itself, but for its results ; and in proportion as the man who desires it possesses a lofty character, his desire for it, being merged in the thought of the uses to which he desires to put it, will itself become equally lofty also. But none the less will the desire of the material wealth form the physical basis in which his loftier desires inhere, just as the impulse of sex remains the physical basis of the deepest and tenderest love which a man feels for a woman, or as the brain is the physical basis of every thought that a man can think. Thus the arguments of the socialists recoil upon their own heads ; and instead of tending to show that the desire of possessing exceptional wealth will ever cease to be indispensable as a motive to exceptional production of it, they have merely succeeded in calling attention to the facts on which the indispensable character of this motive depends.

We have not, however, finished with this question yet. There is a further set of objections still remaining to be considered which, whilst based on an admission that wealth-production is motived by the desire of wealth, aims at showing that this fact does not necessarily result in more than a fraction of the consequences which have up to this time flowed from it, but merely shows in reality that those consequences are unalterable, and adds new force to the arguments that have just been urged with regard to them.

The objections referred to are those embodied in the well-known contention that though the posses- sion of exceptional wealth must be allowed to the exceptional men who are actually engaged in pro- ducing it, and the exercise of whose business ability is just as essential to the country's prosperity as to their own, yet this possession of wealth should be limited to themselves personally, and should not be allowed to distribute itself amongst their idle and inefficient families. In other words, it is urged that whilst the founders and conductors of businesses are entitled to the incomes, no matter how large, that are due to the exercise of their own powers, these incomes should cease with the cessation of the powers that caused them, and should not be allowed to perpetuate themselves, as they do now, in the shape of interest paid to the passive owners of capital. Such an arrangement, it is maintained by those who advocate it, would at once coincide with the dictates of abstract justice, and whilst securing to the exceptional wealth-producer, whose services society requires, the full reward and motive necessary to ensure his activity, would enrich the community at large by distributing amongst it an enormous income, which at present, instead of stimulating anybody to any useful exertion, merely keeps a number of men in idleness. And this contention at first sight does not lack plausibility either in respect of the question of abstract justice which it raises, or of the practical consequences which, according to it, the arrangement in question would produce. When we examine it closely, however, the plausibility vanishes, and abstract justice and practical reason alike condemn the appeals thus made to them as founded entirely on misconception.

Let us deal with the question of abstract justice first. Those who denounce interest or unearned income as unjust, invariably state their case in the following simple form. There are only two ways, they say, in which a man can become possessed of wealth—either by producing such and such an amount himself, or by appropriating such and such an amount that has been produced by another person ; or, as they frequently put it, with an air of solemn sententiousness, "*A man can get an income only by working or by stealing: there is no third way!*" Now one conclusive answer to this puerile, though popular, sophism has, strangely enough, been given by Mr. Henry George, who, though eager to adopt any argument that could be used to assail the rich, was, nevertheless, not taken in by this. Mr. George pointed out that one kind of wealth, at all events,—and we may add that in this we have wealth in its oldest form—consists of possessions which have been neither made by the possessors nor yet stolen by them. That is to say, it consists of flocks and herds. Mr. George pointed out also that whole classes of possessions besides are, for by far the larger part of their value, equally independent of either work or theft. Such posses- This is utterly sions are wines, whose quality improves with time, untrue, as the case of flocks and whose value, consequently, whether in exchange and herds shows us; or use, is increased from year to year by the secret operations of nature. But Mr. George, though his arguments were true so far as they went, did little more than touch the hem of the question; for flocks and herds, and commodities that grow valuable as they mature themselves, form but a small, though they do form a typical, portion of wealth that may come to a man without his having produced it himself, and without his stealing it from any other human producer. And this is the wealth which is actually produced by capital.

In order to show the reader that capital is an but the chief actual producer, in as true a sense as labour is, or producer of wealth that is the ability by which labour is directed, let us begin not worked for is capital, by considering fixed capital as distinct from wage which is past productive capital, and by considering it in its simplest forms. ability stored By fixed capital is meant any tools, machines, or up and externalised. materials by which man's efficiency as a producer of wealth is increased; and we will take as examples of these the three following things—a dart or missile by which game may be killed; a heap of manure by which a peasant's field may be fertilised; and a horse which a peasant uses for ploughing and kindred purposes. Now let us imagine a race of savages who use no missiles at all, but catch their game merely by sleight of hand. If a man is entitled to such game as he catches, the exceptionally dexterous hunter who catches most will be necessarily the rightful possessor of more game than his fellows. This will be granted by those who admit that work constitutes a true, and the only true title to possession.

Such being the case, then, let us alter our supposition somewhat, and suppose that the hunters, instead of catching the game with their hands, kill it with wooden darts; and that the amount of game which each hunter will secure in a day depends not on the skill with which the darts are thrown, but on the skill with which the darts are made. Under these circumstances, the hunter who secures most will not be the man who is quickest in seizing the quarry with his hands, but the man who makes the darts that will reach their mark most certainly; and yet no one would say that he was less entitled to what he took, because his exceptional skill, before it could become effectual, was obliged to become embodied in some object external to himself.

In the same way, if two peasants are cultivating similar fields, and one, by sheer hard work, raises a larger crop than the other, his right to his larger crop would not be denied by anybody. Let us suppose, then, that instead of working harder than his neighbour he works more intelligently, that he saves and stores up as manure materials which his

neighbour wastes ; and that every year, through Book IV Chapter 2
the powers accumulated in his manure heap, he can
raise a larger crop than his neighbour, though he are forms of capital which are actual pro-duce, and the product belongs to those who own them.
actually works less. Would any one affirm that the
man lost his right to his extra produce because he
produced it indirectly by the external agency of
his manure, and not directly by overstraining his
muscles ? Or again, if one of the peasants raised
a larger crop than his neighbour because, whilst his
neighbour spent all his money in drinking, he himself saved it and bought a horse, would any one maintain that the extra crop due to the work which the
horse performed for its owner did not belong to the
owner, but was stolen by him from the other man ?

No one would put forward an argument so absurd The same is the case with such capital as engines and manufactory plant.
as this. And yet the wooden darts of the savage
and the manure heap and the horse of the peasant are
neither more nor less than portions of fixed capital,
just as a steam engine is, or a cotton mill with all its
plant. Fixed capital is merely productive ability
which, instead of acting directly in the production of
goods for the consumer, stores itself up in externalised
means of production, so that it may, with accumulated force, produce such goods indirectly ; and the
additional wealth which a man produces by a new
machine is just as much produced by himself as is
the additional crop which he raises from a patch of
land by the employment of a horse which he has
bought, or manure which he has himself concocted.
Indeed, fixed capital may be compared to a breed
of artificial horses, or if we like the simile better, to

a race of iron slaves. The amount of wealth which the employment of a machine adds to the amount that would be produced without it by a given These implements are like
a race of iron
negroes, and
are producers
as truly as live
negroes would
be. number of labourers, is produced by the machine itself just as truly as it would be if the machine, instead of a structure of wheels and framework, took the form of a gang of artificial negroes, who only betrayed the fact that they were not human by the heat of their breath, an occasional unearthly whistle, and the different language in which they required to have their orders given them. The machine produces this increment, but certain men produced the machine; and therefore the increment is in reality produced by the men, just as truly as when a murdered man has been killed by a bullet from a rifle, his death has been caused by the murderer who aimed and discharged the weapon.

Indirectly,
wage capital is
also a producer
in the same
way. And what is true of fixed capital is true of wage capital also; for fixed capital, such as machines, buildings, or railways, is the result of wage capital, as employed to direct labour, and is therefore wage-capital externalised in the objective results of its employment. But fixed capital, or a man's productive power externalised, differs from his productive power when exercised by himself through wage capital. It is a part of his power which he can separate from his own personality, and which he can make over to others, just as a slave-owner might make over a body of slaves; only these are slaves whose enslavement does them no wrong, and who belong

by right to the men whose enterprise and whose
intellect created them.

Capital, then, as such, is as true a producer of And indeed
till they saw
that this argu-
ment could be
turned against
themselves, it
was strongly
urged by the
socialists. wealth as the men were who in the first instance
produced it ; and when one of them passes a
portion of it on to his son, and with it the income
that results from it, this income is nothing that is
stolen from other men, but is simply a part of the
product produced by the artificial slaves, the use of
whom other men for their own advantage borrow,
and who rightly belong to the lender because he has
received them from his fathers, who created them.
And should any socialist quarrel with this reasoning, it will be sufficient to point out to him that it
is neither more nor less than the reasoning which,
till only a few years ago, the leaders of socialism
themselves were never weary of employing. Capital,
said Lassalle, is merely labour fossilised : and so
long as labour was held to be the only wealth-
producer, the socialists urged that capital belonged
to the labourers, because it represented the labour
of their fathers, whose heirs they were. But with
the gradual disappearance of the doctrine that
labour is the sole producer, it is becoming more and
more evident that capital is not what Lassalle
thought it was—that it is not fossilised labour, but
fossilised business ability. In other words, it does
not, except in its earliest stages, represent on the
part of producers a process of exceptional saving.
What it does represent is a process of exceptional
production. Since then the labourers, as labourers,

would have been the rightful heirs to all capital, if all capital had been produced by the common labour of their parents, those who have actually inherited it must be its rightful owners in fact, because in fact it has been produced by the ability of the exceptional men who left it to them.

But the whole of this argument, based on the claims of abstract justice, would avail very little to defend the income of the mere owner of capital if his position rested upon its abstract justice only, and if his right to his income did not form a part of the very conditions that render the production of wealth possible. The part which the right to income from capital plays when the ownership of the capital is divorced from any active employment of it, depends on the fact that the right to income of this kind is what gives to wealth the larger part of its value, and renders the desire of it efficient as a social motive.

The ways in which it does this are many and various ; and because it is impossible to indicate them in any simple or single formula, certain people may imagine that they have no importance. Such people might as well argue that no complicated process is an important process, or that no results are necessary when many causes combine to produce them.

The most obvious of the reasons why the right to income from capital forms in the eyes of the exceptional wealth-producer a principal element in the desirability of the wealth produced by him has

its root in the facts of family affection. In spite of
the selfishness which distinguishes so much of human
action, a man's desire to secure for his family such is what mainly
wealth as he can is one of the strongest motives of makes men
anxious to pro-
human activity known ; and the fact that it operates duce it ;
in the case of many who are otherwise selfish shows
how deeply it is engrained in the human character.
It may, indeed, be regarded as a kind of selfishness
itself ; and the vigorous and practical men who
have exceptional faculties for wealth-production are
precisely those in whom it is strongest and most
persistent. Men like these would never for a
moment tolerate an arrangement which permitted
the head of the family to keep his wife and children since if income-yielding
in luxury so long as he lived, but would condemn capital could
not be amassed
all of them, the moment he happened to die, to be and bequeathed,
turned by the butler and footmen into the street as wealthy men
could make no
beggars. provision for
It has been said that this family feeling on the their families,
part of the great wealth-producer may be regarded
as a species of selfishness ; and there is nothing very
recondite in the process by which it comes to be so.
Such a man, no matter how selfish, values his family
because it happens to be his own. His own importance is enhanced by the success and brilliancy of
its members ; and the possession of a fashionable
wife, and a popular and well-bred son, reflects
almost as much credit on him as the possession of a
gentleman for his grandfather. For this reason, if
for no others, he will do for them everything that
exceptional wealth will enable him to do. Wealth,

however, depends for its effects on those who enjoy it, not merely on its present enjoyment, but on the prospect of its continued possession ; and unless the man who is making a fortune by his ability may bequeath to one of his children, at all events, a position similar to his own, and something exceptional in the way of wealth to all, the money which he spends on them during his own lifetime will be wasted.   The whole social importance which

nor would
wealth give
pleasure to
those who
might at any
moment be
beggars.
wealth might have given them would be gone.   The tastes and the peculiar cultivation which wealth is capable of securing for those who are from their earliest years surrounded with it, they would under such circumstances neglect to acquire at all ; or, if they did acquire them, they would be living in a fool's paradise, for when their father died, and their wealth consequently vanished, they would be infinitely worse off than those who had never possessed it.   They would resemble nothing so much as plants that had been grown in a conservatory, merely that, when on the point of flowering, they might be bedded out in the frost.

If, then, for the selfish, or even the heartless parent, wealth would in most cases lose the larger part of its attractions unless it could be accumulated and bequeathed to others in the shape of income-yielding property, for the normally affectionate parent its attractions would be reduced yet further.

But the full part which heritable incomes play, in rendering wealth desirable in the eyes of exceptional men, is not to be understood by considering such a man and his family singly.   For the life and the

ambitions of a family are not self-contained.   They

imply and depend upon relations with other families;

and these other families will be valued, and inter-

course with them will be rendered possible, not by the

bare fact that they are the possessors of so much

money, but by the fact that they have the habits and

interests which result, and result only, in the social

atmosphere created by a number of assured incomes,

wholly independent of any daily struggle to make

them.   It is easy to see that no rich society would be

endurable if the only men in it were men who had

just made their fortunes, and if, on their deaths, their

families disappeared from it in the gulfs of destitu-

tion.   Anything more exquisitely ludicrous than the

socialistic proposal that great wealth-producers should

be allowed large incomes to spend, but that they must

not on any account be allowed to invest any part

of them, or use it in a way by which more income

may result from it—anything more ludicrous than

this it is not possible to conceive.   It is—to recur

to an illustration used already—like proposing that

a peasant who is more industrious than his neigh-

bours shall be allowed all the money which the sale

of his extra produce brings him, provided only that

he spends it on brandy, or beer, or absinthe; but

that if he save it up and buys a useful horse with

it, his purchase shall be confiscated by the State,

because a horse is productive capital.   This pro-

posal, however, is not only ludicrous in theory, but

it would, if put into practice, result in a sort of

Book IV
Chapter 2

Moreover, if
incomes were
not heritable,
wealth could
produce none
of those social
results, such as
continuous
culture, etc.,
which make it
valuable.

The wealth
that ceased
with the men
that actually
made it would
produce a
society of
beasts.

society more vile and bestial than anything which the world has ever known. For the sole advantage which in that case wealth would bring to its producer would consist in the meat and drink and other means of physical pleasure which he and his family could consume or enjoy during his lifetime—before he retired to the grave, and his wife and children to the workhouse.

Wealth is desirable because it is the physical basis of an enlarged life ;

The main value of wealth in the eyes of the great wealth-producer does not consist in its ministering to brief spasms of self-indulgence, but in the fact of its being the foundation of an equable and sustained life, in which the physical pleasures are refined rather than intensified, and the time employed by the majority in producing the necessaries of existence is given not to sloth, but to other kinds of exertion. A life of this kind is impossible except in a society of which a large section not only

and there must thus be continuity in the possession of wealth.

possesses wealth, but is accustomed to its possession, and is characterised by accomplishments, tastes, principles, and kinds of knowledge, which can be developed and acquired only when the continuance of its possession is assured. In other words, those men on whose exceptional business ability the productive processes of the entire community depend, and who are the cause of growth in the incomes of the mass of the community, just as truly as they are the producers of their own fortunes, are motived to activity less by the desire of the wealth which comes to them day by day through their own direct exertions, and which would cease instantly

when these exertions were suspended, than they
are by the desire of wealth that shall come to them indirectly, not as the product of their exceptional exertions in the present, but as the product of the accumulated product of their exceptional exertions in the past—the product of those stored-up forces with which they have enriched the world, and which, whilst rendering help to thousands of men besides, will continue to render a tribute to their creators and their creators' children.

Thus, to express the matter in brief and familiar language, the sustained development and exercise of exceptional ability in wealth-production implies the possession by those who monopolise this ability, not merely of that portion of those products which are called the wages of superintendence, but also to that portion which is called interest on capital. For just as the control of capital affords the only means by which, under free institutions, the great man can apply his faculties so as to increase the production of wealth, so does the right to interest, or to the products of the capital accumulated by him, constitute the chief reward by the desire of which the exercise of his faculties is stimulated.

There is a further point, however, which now remains to be noticed. When it is said that the great wealth-producer is motived mainly by the desire to enjoy an amount of wealth proportionate to what is produced by him, it is not asserted that in order to gratify this desire it is necessary that he should be able to appropriate the whole of what

Hence the great wealth-producer demands the possession not only of what he produces directly, but of what he produces indirectly through his past products.

21

The majority not only may, but do acquire a share of the increment produced by the great man ;

is produced by him. On the contrary, of that constantly growing product which is added by the great man's faculties to the product of ordinary labour, and out of which the income of the great man comes, a portion is capable of being appropriated by the ordinary labourers themselves. Indeed, the masses of the community are partakers in material progress, and have an interest in material progress solely because, as an actual fact, a considerable percentage of this added product goes to them ; and though few of our so-called "labour leaders" recognise this truth, all the hopes of enrichment which they hold out to their followers imply nothing whatever beyond the securing a larger amount of an increment which is produced not by themselves but others. An important question, therefore, arises in this way as to how far the product of the great men can be taxed and handed over as a bonus to average labour without weakening the

but whatever this share may be, it can never be such as to render social conditions equal.

motives which prompt the great men to produce it. This is a question to which, by à priori reasoning, it is absolutely impossible to give any definite answer. It is a question that can be solved only by cautious practical experiment ; and the answer will vary constantly with times, places, and circumstances. All that can be asserted here, and it is all that requires to be insisted on, is that the amount of wealth which the exceptional wealth-producer can secure must be proportionate to what is produced by him, however far short of the whole of it ; and that it must not be diminished to such an extent as

will render it less exceptional as the object of an
ambitious and strenuous man's desire.

In other words, that graduation of social circumstances, those differences in ways of living, in habits,
manners, accomplishments, and social functions,
which have their physical basis in varying degrees
of wealth, and give to civilised society what is its
present, as it has been its past character—these
graduations of social circumstances, which it is the
cherished dream of the socialists to do away with,
are indestructible so long as civilisation lasts. If
they perish, civilisation will perish also ; when civilisation is restored they will reappear along with it ;
and however they may be modified or adjusted,
they can never be even approximately effaced.

It is the facts briefly indicated in the present
chapter which the socialists of to-day are principally
distinguished by ignoring ; and it is these facts
which render socialism for ever impossible.

This truth, when once generally recognised, will
lead to many practical consequences, of which the
most immediately important will be dealt with in
the following chapter.

CHAPTER III

THE two great facts, then, that have been elucidated by our inquiry thus far, are these : in the first place, all progress and civilisation, and more especially all production of wealth, results from a complicated process in which, man for man, a minority plays a part incalculably greater than the majority; and consequently, in the second place, the minority, man for man, possesses wealth that is correspondingly greater than the wealth of the majority, likewise. In addition to these facts a third has been elucidated also, to which it is desirable that we should give renewed attention. Since great men not only pro-

The wealthy class, owing to inheritance, is always much more numerous than the great men actually engaged at any given time in production.

duce wealth directly, but produce it indirectly by producing wealth which produces it, and which they are enabled to hand on to their children, the wealthy class is at any particular moment always more numerous than those members of it who are engaged actually in production. In Great Britain, for example, it has been estimated that two-thirds of the aggregate income that pays income tax is rent or interest on capital, and that one-third represents the direct products of work. We may therefore here adopt the rough hypothesis that out of each generation of our wealthy class a third part is enriching itself by the process of direct production, and two-thirds are living on the products produced for them indirectly by the capital or the means of production which were created by their fathers and their grandfathers. Now such being the case, what we have to notice is as follows. Though the members of the wealthy class are not always changing, as they would be were no saving of capital, no interest, and no bequest allowed, they are still changing gradually from generation to generation, so that whilst the class, as a class, always possesses a nucleus of families with whom wealth and the traditions of wealth are hereditary, a number of individuals born in it are constantly disappearing over its borders, and a number of other individuals are constantly passing into it.[1]

But though inheritance gives a certain permanence to the wealthy class, the families belonging to it are constantly, if slowly, changing,.

[1] The most permanent form of hereditary wealth is land ; but only a small minority of our existing landed families existed as landed families at the time of the last Heralds' visitation. Thus, though the estates of this country are as old as the country itself, the actual possession of a large proportion of them by their owners, at any given time, represents their purchase by wealth recently created, and is, in fact, recent wealth converted into another form.

And if there is a change like this in the possession of landed wealth, there is a still more rapid change in the possession of commercial capital. One of the many childish assumptions of Karl Marx was the assumption on which a good deal of his reasoning rests—that the English middle classes of the present century owed their capital and positions to social opportunities which had come to them as the heirs and descendants of the merchants and wealthier

Thus in spite of the permanence which interest gives to wealth, the families that live merely on interest are constantly tending to disappear, and their places are being taken by the men whose exceptional faculties, whose business ability, whose enterprise and strenuous will, actually contribute most to the productive forces of the country. It was observed by J. S. Mill with regard to political government that this "is always in, *or is passing into*, the hands" of the men who are at the time the true repositories of power. In the same way the wealth of any progressive country is always in, or is passing into, the hands of the men who by their own abilities are engaged actively in producing it.

and new men
are constantly
forcing their
way into it.

sheep-farmers who began to make fortunes four hundred years ago. As a matter of fact by far the larger part of the great commercial businesses and commercial fortunes now existing in this country have been founded during the past hundred, and many within the past fifty years, by men who were the sons of ordinary wage-paid labourers, and who were no more heirs to the men who formed the middle class under the Tudors than they were to the merchants who are celebrated in the *Arabian Nights*. That such is the case is shown with sufficient clearness by the following figures, which refer to commercial incomes during the thirty years which followed the first Great Exhibition. During these years, whilst the population increased by about 30 per cent, fortunes of over ten thousand a year were multiplied by 100 per cent, fortunes of from five to ten thousand by 96 per cent, and fortunes of from five to six hundred by 308 per cent. It is obvious, then, that when a class is augmented in one generation by a number of new members from three to ten times as great as its natural increase would account for, most of its new members must have come to it from some class outside, and have gained their place in it solely by their own exertions.

Such being the case, then, the material civilisa- tion of a country—the wealth of the few or the pro- gressive comfort of the many—depends on the ex- tent to which its potentially great wealth-producers, as they come into the world, generation after genera- tion, are induced by circumstances to develop their exceptional talents, and devote them to the main- tenance and improvement of the productive process. For those, therefore, who regard the material wel- fare of a community as the test and basis of its welfare in all other ways, the abiding social problem is always this : how to adjust circumstances in such a way that the smallest possible number of these potentially great wealth-producers may be wasted, and the largest possible number may be induced to exert themselves to the utmost.

Indeed the wealth of the country de- pends on the men potentially great as pro- ducers actual- ising their talents and producing the wealth that raises them.

One set of conditions essential to this result has been described already—those, that is to say, by which the possession of wealth is secured to the producers of it, and the persons to whom they leave it. But to these must be added another set of an entirely distinct character—that is to say, the con- ditions which, the motive to exertion being given, shall render exertion of the kind required possible for the largest number who happen to be theoretically capable of it. Now modern democratic thinkers have supplied the world with a formula by which, in their judgment, these conditions are sufficiently indicated. This formula is "equality of oppor- tunity," and we cannot begin our consideration of the question better than by taking this as a starting-

It is therefore obvious that wealth will increase in pro- portion as these potentially great men have the opportunity of actualising their produc- tive powers.

point, and asking what truth is contained in it.   We may at once admit, then, that if it is taken in an abstract sense, it sums up a truth which is, beyond doubt, indisputable ; for if each individual having exceptional potentialities as a wealth - producer, which require nothing but the favour of circumstances to ensure their being turned into actualities, could be provided with circumstances so nicely adapted to his idiosyncrasies that these potentialities might be developed to the utmost extent possible, the productive powers of the community, it is almost needless to observe, would be raised in that case to their utmost possible efficiency.   Such an ideal condition of things as this, however, is impossible for the following, if for no other, reason.   Successful parents as a rule will employ part of their wealth— at all events they will employ the positions which they have won by their own ability—to provide opportunities of a special kind for their sons ; therefore, whatever the State might do for its youths and young men in general, exceptional parents for their sons would be able to do something more. Equality of opportunity, therefore, represents an ideal condition which we never can reach, but to which we can only approximate ; and the only practical questions for us are accordingly these : how far towards this ideal can political action carry us, and what results are to be anticipated from our nearest possible approach to it ?

Now the answer to both these questions will very largely depend on the existing conditions of the community with reference to which they are asked. For though men's powers of equalising opportunities are limited, their powers of making them unequal may be said to be indefinitely great ; and the more unequal they have been made at the time when we ask our questions, the greater will the progress be which there will be room for us to make towards equalising them, and the greater will be the social advantages which we may hope to secure by making it. In France, for example, before the first Revolution, the laws affecting industry had almost ruined the nation, not because by unduly favouring one class they led to wealth being concentrated, but because by unduly hampering other classes they prevented its being produced ; and the sweeping away by the Revolution of the old feudal inequalities, though it had none of the millennial effects which the Revolutionists themselves hoped for, has had others equally striking, though of a very different kind. It has not made men equal in point of wealth, but it has increased to an astonishing extent the wealth of all classes alike. And the way in which it has done this has been by removing artificial impediments to the development and free exercise of exceptional productive talent ; or in other words, by an equalisation of economic opportunities.

But the kind of equality that has thus been reached may be described as being of a negative rather than a positive kind. It depends on the absence of artificial impediments to production, rather than on the supply of any artificial helps to

Side notes: In a country where opportunities have been made artificially unequal there will be room for a great deal of equalisation.

But removing artificial impediments is only a negative kind of equalisation.

it ; which means that it depends on the absence of everything that might obstruct the strong, rather than on measures or institutions that should artificially lend strength to the weak.   Now, so far as industrial ability of the highest kind is concerned, it is probable that this negative condition of things, which is merely the complete embodiment of a policy of *laisser-faire*, represents the utmost that, in any civilised country, can be done by the process of equalisation with any beneficial result.   For in wealth-production the men whose capacities are

It is probable,
however, that
for the develop-
ment of genius
of the highest
order this is
all that is
needful, really of the first order will, when not positively impeded, make their own opportunities for themselves ; and the genius who is born with every opportunity waiting for him has but a few years' start of the genius who is born with none.   That such is the case is abundantly illustrated by history. If we consider the most famous of the men whose originality of mind and extraordinary spirit of enterprise have been chief amongst the forces which have enriched the civilised world, we shall find that those whose names most readily occur to us have had no opportunities save such as their own genius made for them.   Arkwright, Cartwright, Watt, Stephenson, the intrepid and enduring adventurers who, in the teeth of prolonged opposition, laid the foundations of the modern manufacture of iron ; Columbus, who gave to Europe a new hemisphere—all these have been men born amongst social circumstances which conspired to deny them rather than to provide them with opportunities.   And if we turn from

Europe to new countries like America, and consider the leaders of economic production there, we shall find that the histories of these men have been similar. Nor, indeed, in this fact is there anything to be wondered at. In the sphere of industry, just as in the sphere of art, the greatest men will never be suppressed. They are always sure to assert themselves, and the struggle with adverse circumstances will, instead of crushing, strengthen them.

It may therefore be safely said that no equalisation of opportunity which goes beyond the abolition of arbitrary and unequal impediments would tend to increase the number of those exceptional men whose produc- *and will secure* tive faculties are really of the first order. And this *the develop- ment of all the* inference is supported by a large number of analogies *genius of the highest kind* drawn from domains of activity other than economic. *that exists.* Any workman's boy, for example, who has any taste for books has now in England, before he is fifteen, more educational opportunities than Shakespeare had in all his lifetime. But the number of Shakespeares has not appreciably increased. Again, popular education has given to the whole French army advantages confined to a few at the time of Napoleon's boyhood. Every private carries the marshal's *bâton* in his knapsack. And yet democratic France, with all its equalisation of opportunity, has not produced a series of new Napoleons. On the contrary, the mountain, after years and generations of labour, does nothing at last but give birth to a Boulanger.

Though faculties of the first order, however, are

independent of artificial assistance, many of an inferior, but still of an exceptional kind, are not ; and
But genius of
a lesser kind,
which would
else be lost,
may no doubt
be elicited by
positive educa-
tional help
from the State ;
it cannot be doubted that the supply. of these last
will depend very largely on the degree to which
facilities for self-development are given by the State
to those who desire to take advantage of them.
Thus, though the spread of education in this country
has not increased the number of Shakespeares, it has
enormously increased the number of those who can
write good English.   And no doubt in the domain
of wealth-production it has had an analogous effect.
This effect, however, though real, has been enormously exaggerated ; and it has been exaggerated for a particular reason.   Social reformers have
confused two things together.   They have confused
talents which are exceptional in their very nature,
though the
amount of such
genius is over-
estimated by
reformers,
because they
confuse talents
rare in them-
selves with
accomplish-
ments that are
only rare
accidentally.
with accomplishments which are exceptional only
because they are not universally taught.   Thus reading and writing, for instance, were rare accomplishments once.   Of all accomplishments they are the
most universal now ; and there is not the least doubt
that there are very many others which, with equal
opportunities, might be acquired by almost anybody,
but which yet, as a matter of fact, are still confined
to a minority.   In this fact that education may increase the accomplishments of a community, social
reformers have fancied that they discovered an indication of the extent to which education could elicit
exceptional talent.   But to call into practical activity
by means of external help exceptional faculties, of
which the supply is necessarily limited, is a very

different process from evoking by similar means
faculties which are potential in everybody, and the
supply of which can be increased indefinitely ; and   The latter can
it is a process, moreover, which produces very   be increased
indefinitely, the
different results.   Let us consider how this is.   former not.

For productive faculties of the highest order,   For real pro-
which not only minister to progress, but initiate it,   ductive genius
there is always
and which make, as if by a conjuring trick, the hands   room,
of the average labourer produce new commodities
of which he never would have dreamed himself—for
faculties such as these, the demand is always un-
limited.   There are productive faculties also, exceptional although they are inferior, the demand for
which is usually greater than the supply.   But with
regard to those faculties or accomplishments which
are only exceptional accidentally, and which might
be, like reading, conceivably made universal, the
case is precisely opposite, and it is so for two reasons.
In the first place, these accomplishments, which
anybody might conceivably acquire—knowledge of
French, for instance, or of book-keeping—though
they may minister to the business of wealth-produc-   but the
economic
tion, yet have no tendency in themselves to make   utility of mere
the business grow.   The number of persons, then,   accomplishments is limited
possessing these accomplishments who at any given   by the conditions of pro-
time can put them to a productive use is limited by   duction at the
time.
the condition in which production at that time is.
Thus the number of clerks which a mercantile firm
can employ is limited by the business which the
firm happens to be doing ; and though this business
might be enlarged by the enterprise of one new

partner, it would not be enlarged, when there were no letters to copy, by the accession of ten young men who could copy letters beautifully. In the second place, even at times when the national business is growing, and the demand for these accomplishments is for the moment greater than the supply, any attempt by the State to make their development general would produce a supply indefinitely greater than the demand. Thus to multiply the number of labourers' sons possessing accomplishments that would fit them for the work of clerks would not be to increase the number of young men who would wear black coats, and sit on stools in offices, instead of working in factories, or laying bricks, or plough-ing. Instead of raising the position of the plough-boy to the same level as the clerk's, it would lower the clerk's salary to the level of the plough-boy's wages ; and clerk and plough-boy would be alike sufferers by the process.

Thus to pro-
duce more
possible clerks
than are
wanted merely
lowers the
wages of those
employed,
without in-
creasing the
utility of those
who are not
employed.

The beneficial effects, then, to be looked for from an equalisation of opportunity have been exaggerated by democratic thinkers because they have failed to perceive those facts. They have confounded the development of accomplishments which might conceivably .be acquired by all with the development of faculties which, even potentially, are possessed by a few only. They see that education can increase the number of possible clerks, and they have therefore imagined that it can, with similar ease and certainty, increase the number of efficient men of genius. It must, however, be distinctly stated that the error in their conclusion is one of exaggeration only. There is much exceptional talent which, though not of the highest order, will, when opportunity is given it, increase the wealth of the community, but which will, without the educational help of the State, be lost; and it may frankly be admitted that, within certain limits, the equalising of educational opportunity plays a very important part in supplying the community with exceptionally efficient citizens.

Book IV Chapter 3

Still, within limits, educational help from the State does much to increase the supply of exceptional, though not great, talent.

But the main difficulties involved in the artificial equalisation of opportunity are not concerned with the problem of how to produce good results by it. They are connected with the problem of how to avoid producing bad results. Let us consider what the possible bad results of it are.

In a general way they are indicated, or indirectly implied, in the saying so dear to the sterner and more thoughtless of the Conservatives—that popular education does nothing but promote discontent. Sweeping statements of this kind, however, though they may have an element of truth in them, are valueless till they have been carefully qualified ; for what we have to ask about them is not whether they are true, but how far they are true, and in what precise senses. Thus, though it is true that the danger of diffusing education lies in the discontent that may thereby be promoted, some kinds of discontent are not dangerous—they are beneficial ; therefore the danger of diffusing education lies in its tendency to promote not discontent generally, but discontent of certain special kinds ; and it is necessary to discriminate carefully what these kinds are.

Now the kind of discontent which Conservatives generally have in view, when they denounce education because they think it tends to promote it, is by no means that from which danger really arises. What they generally have in view is a discontent with his circumstances which they think education will produce in the average working man. In reality, however, the primary danger of education is not to be looked for in its effects upon average men at all. It is to be looked for in its effects upon men who are distinctly exceptional.

In order to understand how this is, let the reader reflect once more on one of the main truths that have been insisted on in the present volume—namely, that though all progress is the work of great or exceptional men, all great or exceptional men do not promote progress equally, and some of them indeed do not promote it at all. Progress results from the victory of the fittest of these over the less fit in the struggle to gain dominion over the thoughts and actions of others. Let the reader reflect also on the analysis that was given of the various qualities which go to make up greatness—that is to say, the qualities by which dominion over others is obtained. It was pointed out that greatness is a highly composite thing ; that it need not necessarily imply any moral, nor indeed any intellectual superiority ; and as an illustration of this it was mentioned that many most

important political movements have been produced
by men whose greatness consisted merely in ordinary sense joined to, and made efficient by, an extraordinary strength of will. It is necessary now to *but whose exceptional gifts are ill-balanced or have some flaw in them.* follow this line of observation farther, and to point out that if extraordinary strength of will can produce beneficial effects when allied with ordinary sense, it is equally capable of producing effects that are mischievous when allied with stupidity, or with that kind of imperfect intellect which is as quick in defending and popularising, as it is in being duped by fallacies. And with these latter qualities it is allied as often as with the former. It is a great mistake to suppose that even the most false and foolish opinions which have influenced multitudes to their own detriment have been originated and promulgated by men who were altogether weak and inferior. On the contrary, most of the follies which have disturbed or retarded civilisation have been due to the influence of men who, though morally or intellectually contemptible, have possessed a vigour of character far beyond what is ordinary.

Now, if education has the effect attributed to it *For if education sets free and stimulates sound intellectual powers,* of liberating the will and developing the intellectual powers of men in whom the intellect is really acute and sound, there is an obvious danger of its having the same effect on men whose intellect is unbalanced and imperfect. To some of such intellects, no doubt, it may give clearness and equilibrium ; but there are *it will similarly stimulate intellects that are not sound,* others for which it does nothing, except to increase their powers of reasoning wrongly ; and when an

22

intellect of this kind is allied with a naturally strong will, the effect of education is to let loose a wild horse, merely in order that it may run away with a lunatic.

It must be remembered that the strength of a man's will, though depending as a potentiality on the character with which he happens to be born, depends as an actual force on his desire for certain objects or results, coupled with the belief that he can attain these by action. Now, when a man's powers of action are capable of realising his desires —as when a man who desires to be wealthy has the talents that produce wealth, or when the man who desires to be Prime Minister has the talents of a great statesman—his career satisfies himself, and is presumably serviceable to his country. In many cases, however, desire is exceptionally great, and generates also a strong impulse to act, but the capacity for that kind of action by which the desired object might be obtained is small. Thus many men desire exceptional wealth, but find themselves incapable of the peculiar kind of action that produces it. Their will, accordingly, if it makes them act at all, is like a steam-engine which merely puts useless machinery into motion; or if it fails to make them act, as it very often does, it shakes them to pieces with a kind of intellectual retching. These unhappy persons owe the condition in which they find them-

selves mainly to an over-estimate of their own powers; and this over-estimate is generally the direct result of education, which, by making them

falsely imagine themselves capable of attaining
wealth, actualises a fruitless desire for it, which might otherwise have remained latent. When education has this effect on a man it is an unmitigated evil for himself, and very frequently for others.

Again, education, besides actualising exceptional desires which are wholly unaccompanied by any exceptional faculties that correspond to them, actualises desires accompanied by faculties which are really exceptional, and which produce results undoubtedly more than ordinary, but are nevertheless incapable of complete development. Many men, for instance, have gifts for music and poetry which, though genuine so far as they go, have yet some fatal defect in them, and will never produce, however devotedly they are exercised, any results possessing artistic value. Now the fact that progress is caused by a struggle between exceptional men, of course implies that some of them shall be less efficient than the others. It is by struggling with the less efficient that the superiority of the most efficient is realised ; and in order that it may be found who the most efficient are, the inferior as well as the superior must put their capacities to the test. It is therefore unavoidably one object of education to stimulate the activity of some exceptional men whose own efforts are foredoomed to ultimate failure. Failures, however, differ in degree and kind. Some men fail because they can accomplish nothing of what they attempt, like the dreamers who have wasted their

Education, again, stimulates faculties that can really produce exceptional results, but not results that are complete.

The progressive struggle requires that

lives in trying to make perpetual motions. Some fail because, though they accomplish something, others accomplish more ; and the production of what is the best makes the second best valueless. Thus nine inventors might produce nine motor-cars, each of which worked well enough to command a considerable sale ; but if a tenth inventor was to produce another which was faster, simpler, more durable, and cheaper than any of these, all the rest would drop out of use altogether, and be practically as valueless as the mad aggregation of wheels by which the seeker for the perpetual motion endeavoured to accomplish the impossible. Between the men who fail, however, because they succeed less than others, and the men who fail because they do not succeed at all, there is a great practical difference. The men who fail only because others succeed better

But those
failures that
promote pro-
gress are fail-
ures that
partially
succeed.
than they do, contribute to the very success of the men by whom they are defeated ; for they raise the standard of achievement which these men have to overpass. But the men who fail because they accomplish nothing waste their own lives without benefiting anybody. In the domain of economic production the truth of this is obvious. It is not less so in the domain of speculative thought. Scientific theories are constantly put forward which, though not true, are sufficiently near the truth to have some definite relation to it ; and those who actually reach it find in errors of this kind an indispensable assistance. Nothing gives to truth so keen and clear an outline as the refuted errors of really powerful thinkers. But

there are errors, on the other hand, which, though
it may be necessary to refute them because they have imposed themselves on a number of ignorant people, do nothing to advance the discovery of truth whatever, and the activity of those who originate them is altogether mischievous. Thus whilst the reasonings of heretical thinkers like Arius, by the controversy they provoked, were very largely instrumental in advancing orthodox theology to really logical completeness, the philosophy of religion owes absolutely nothing to Joanna Southcott or the American prophet Harris. Accordingly, whilst it is impossible to say with precision where the line is to be drawn between the exceptional talents which, if developed, would be of use in the progressive struggle and those 'which are so defective that their influences would be merely mischievous, it is obvious that talent of this latter kind is sufficiently plentiful to render its development dangerous.

History teems with examples of this fact, and so do the unwritten annals of the social life around us. Henri Murger in his studies of Bohemian Paris bears eloquent witness to the tragic absurdity of the results caused by the development of imperfect artistic talent, and the miserable endings of men who, if they had not tried to be artists, might have lived and thriven as honest and healthy *ouvriers;* whilst, according as we hold vaccination to be a blessing to the world or a curse, we must necessarily hold that it would have been far better for everybody if the talents of the men who invented it, or else those of the men who now oppose it, had been killed by the frosts of ignorance, and never allowed to blossom.

But the
commonest
example of this
kind of man is
the socialistic
agitator,   But the commonest examples of talent that is wholly mischievous are afforded by certain classes of politicians and social agitators. There is a large number of men whose potential activity is considerable, and whose intellect has a natural nimbleness which will enable them, when stimulated by education, to seize on plausible fallacies and impose them both on themselves and others. Politicians of this class are familiar figures enough. The social agitator, whose mental equipment is similar, is more familiar still. Many attempts have been made to give a scientific explanation of those constant attacks on the existing organisation of society which are common to all civilised countries, and go by the name of socialism. Socialism is said by some to be the protest of increasing poverty against increasing wealth ; by some to be the natural voice of highly organised labour, which has come at last to be capable of self-government ; and by some to be an embodiment of the esoteric philosophy of Hegel. In reality it is the embodiment of the results of indiscriminate education on talents which are exceptional, but at the same time inefficient. The avowed object of socialism is a redistribution of wealth ; but the most striking characteristic of all the socialistic leaders has been an incapacity to produce the thing which they are so anxious to distribute. The wish to re-

distribute it in some of them arises from sentiments
of benevolence ; in some from fallacious reasoning ; and in some from personal envy ; but in none has it been accompanied by those particular faculties on which the actual production of wealth in large quantities depends. Socialism, therefore, so far as it is who demands the re-distribution of wealth whilst absolutely powerless to produce it, a serious theory, is essentially an attempt on the part of men who are themselves economically impotent to prove that they, and others like them, have some reasonable right to possess and divide amongst themselves what they are constitutionally powerless to make for themselves. The result has been the elaboration of a theory of production which sometimes declares that wealth is produced by "aggregates of conditions," or "social inheritances," or "environments," as Mr. Spencer, Mr. Bellamy, and Mr. Sidney Webb tell us ; and sometimes that it and who consequently invents false theories about its production, which do nothing but demoralise those who are duped by them. is produced by "average labour measured by time," as Karl Marx tells us,—the one doctrine being that wealth is produced by nobody, and that one man has thus as good a right to it as another ; the other being that it is produced in equal quantities by everybody, and that everybody on that ground has a right to an equal quantity of it. Both doctrines agree in this, that they altogether miss and divert the attention of the mind from the forces and conditions on which wealth-production depends in reality.

Now if the elaboration of these fallacies had been confined to men who were capable of presenting them in a really arguable form, and if they had been (though even these theories can be dis- promulgated only amongst classes who were capable

Book IV
Chapter 3

cussed with
profit under
certain circum-
stances).
of passing a scientific judgment on them, they might have played—and within limits they have played—a valuable part in eliciting the truth opposed to them. But they have become wholly mischievous when, through the agency of indiscriminate education, they have influenced men who, whilst wanting in intellectual judgment, are nevertheless endowed with a potential activity of character, and who, when this is developed, at once become powerful agents in disseminating fallacies amongst others even less capable of criticising them than themselves. Thus many of the leaders of the "new unionism" in England are to be credited with energy of a really remarkable kind; but unfortunately the energy is united to such defective intellectual powers, that the more vigorously these are employed, the more mischievous and absurd is the result. The general resolutions that have been passed at Trade Union conferences declaring that no progress is possible till all the means of production shall have been nationalised, or the doctrine of the "new unionists" that wages control prices, are all results of the exercise of faculties which, though in some respects doubtless superior to those of the average man, had far better have never been developed at all.

Men like these
embody the
two chief
dangers of the
artificial equal-
isation of
educational
opportunity,
It is men like these—the men with ill-balanced or abortive talents—the men with strong wills and defective intellects, the men whose ambition is developed by the smallest educational stimulus, but who have no talents proportionate to it which any

education could develop—it is men like these who invest with its principal dangers the equalisation of educational opportunity ; and if education, as so many Conservatives say, really does nothing but promote popular discontent, it promotes discontent amongst the great masses of the population less from the manner in which it affects the average man directly, than from the manner in which it affects men who are inefficiently exceptional, and who, not having the gifts that would enable them to rise in any society, endeavour to persuade the masses that society, as at present constituted, is an organised conspiracy of the few to keep everybody else down.

The equalisation of educational opportunity has, therefore, two dangers—the danger of developing wants in the average man which could never be generally satisfied under any social arrangements ; and the danger of developing the talents of a certain class of exceptional men which are naturally incomplete, and which the more fully they were developed, would only become more mischievous both to their possessors and to society.

And these dangers correspond with the two objects for the sake of which the equalisation of educational opportunity is advocated. One of these objects is the raising the condition of the average man ; the other is the securing, alike for himself and for society, the full benefit of the potential gifts of the exceptional man. The average man, however, is not made better or happier by being filled in early life with importunate wants and propensities which he will, when he comes to maturity, be unable to gratify; nor is any one made better or happier by the development of gifts which, however exceptional, can, by reason of their incompleteness, do nothing but give currency to error, or initiate abortive action.

It is the latter of these dangers that is practically the source of the former. The average man would, as has been said already, probably suffer little from over-development under existing systems of education if it were not for the effects of these systems on inefficiently exceptional men whose superiorities ought never to be developed at all. It is doubtless impossible to avoid this danger completely. If educational opportunities are to be of a kind that will enable the efficiently exceptional to work their way to the top, and advance or maintain civilisation by their influence or domination over others, it is inevitable that a certain proportion of the inefficiently exceptional will be induced to develop their unhappy capabilities also; but the number of these may, at all events, be reduced to a minimum. The fundamental fault of contemporary educational theories is, that in proportion to the completeness with which they were carried out, they would tend to raise the number of these men to a maximum. And the reason why they would have this tendency is that they are founded on two absolutely false principles.

The first of these principles is, that whatever potential talents any man may possess, it is desirable to assist and encourage him to develop them to the utmost. The second is that the type of education and culture to which education generally should, so far as is possible, be assimilated, is the kind of education and culture that is actually prevalent amongst the rich.

It is impossible to meet these principles with too emphatic a negative.

The first of them is false because, as has just been shown, there is a large amount of really exceptional talent which, if developed, would work nothing but mischief, and which ought, consequently, for the sake of everybody, not to be developed, but suppressed. The second is false so is the theory because all tastes and talents are good or bad, that all tastes should be useful for a man or useless, according to the cultivated in all alike. The conditions under which his life will be passed ; and education proper for the the conditions of the rich are altogether exceptional. rich is not a Societies have existed in which they have been type but an exception. enjoyed by nobody. It would be impossible to construct a society in which they should be enjoyed by more than a few. The attempt, therefore, to give to everybody a rich man's education is like including skating in the curriculum, and fur coats in the wardrobe, of a thousand boys, when nine hundred of them are to spend their lives in the tropics.

Both these false principles rest on that radically These false false theory of society which it is the principal object the false belief of the present volume to expose—the theory that that equal education civilisation is the product of men approximately could ever produce equal equal in capacities, and that in proportion as these social conditions. equal capacities have equal opportunities of development, there will naturally be an approximation to an equality of social conditions. The facts of the case are precisely the reverse of these. Civilisation originated in, and is still maintained by, men whose capacities are unequal to those of the majority ; and just as there is no tendency towards equality in capacity, so, for reasons which have been explained in the last chapter, there is no tendency towards equality in social conditions. Inequalities of condition may at some times be greater than at others, but the fact that at times they show a tendency to become less is no more a sign that they have any tendency to disappear than the fact that an economy has been effected in the consumption of coal on board a steamship is a sign that steam has a tendency to be generated without fire. It is therefore a scientific certainty that of each generation of children in every civilised country the majority will, throughout their subsequent lives, occupy positions very different

The majority of each class will remain in the class in which they were born.
from those of the few. Most of the members of each class will remain in the position in which they were born ; but there will be a gradual descent from the upper classes of their weaker members into the lower, and amongst the stronger members of the lower classes there will be a constant potential desire

Only the efficiently exceptional can rise out of their own class ;
to push their way into the upper. Some of these last are strong in potential desire only. With others the strength of desire is accompanied by corresponding talent, by means of which, if developed, the position which they desire will be obtained. It will be obtained by the talent of these men, because the talent of such men is creative ; and when it is

developed it renders those who possess it actual additions to the civilising forces of the community.

With regard, then, to exceptional men, the object of education should be to stimulate the ambitions of those of them whose talents are efficient, whilst discouraging the ambitions of those whose talents are inherently defective. The stronger the ambitions of the former are, the better for themselves and for the community. Men like these are the true gold-mines of their country. The stronger the ambitions and the larger the opportunities of the latter, the more will the health and strength of the social organism be interfered with.

and it is the ambition of the efficiently exceptional only that it is really desirable to stimulate.

With regard to the average man, the object of education should be to develop in him such tastes or accomplishments as will assist him in the work by which he is to live, and enable him to make the most of such means of enjoyment as are within his reach, whilst leaving him untormented with a desire for enjoyments that are beyond it ; and the crucial fact on which it is necessary to insist is that the circumstances of different classes are permanently and necessarily different, and that for the average man of each class the education that will make the most of his life is necessarily different also.

The average man should be taught to aim at embellishing his position, not at escaping from it.

In other words, the only true equality of educational opportunity is an equal opportunity for each, not of acquiring the same knowledge or developing the same faculties, but of acquiring the knowledge and of developing the faculties which, given his circumstances and given his natural capacities, will do most to make him a useful, a contented, and a happy man.

Unfortunately these conclusions, simple and obvious as they seem, run directly counter to that entire theory of society which, with more or less consciousness, and with more or less precision, is held by the school of writers, reformers, and politicians, who suppose themselves, in some exclusive sense, to have social progress at heart; and also to that mass of diffused sentiment which, though not expressing itself formally in any theoretical propositions, has that theory as its foundation, and bears to it, as a political force, the same relation that vapour bears to water. These conclusions, therefore, which imply inequality in capacity as the cause of social progress, and inequality in social circumstances as the necessary and permanent conditions of it, are, like most of the other conclusions put forward in this work, certain to be met with objections of the most vehement kind, which it will now be necessary for us fairly and carefully to consider. We shall find that, as we do so, the entire arguments of the present work are summed up and brought together before us; and however incompatible they may be with the false conception of progress, of class relationships, and of the structure of society generally, which are at present mischievously popular, they form the foundation of hopes, for all classes, far more solid than those, the fallacy of which they aim at demonstrating.

CHAPTER IV

MAN does not live by wealth alone, and progress is not concerned solely with the production and the distribution of it. But the processes involved in the production and distribution of wealth, though far from being coextensive with all social progress, are typical of it. They form, moreover, the subject with regard to which contending politicians and reformers practically join issue; and it is mainly because inequality in the possession of wealth is affirmed to be a permanent and necessary feature of civilisation, that the conclusions here put forward will be attacked. *The radical politician will object to the foregoing conclusions in terms with which we are familiar.*

The objections that will be brought against them will take two forms ; one being the form which will be given them by the radical or socialistic politician ; the other the form which will be given to them by the radical or socialistic theorist.

The radical or socialistic politician, whether he is journalist or popular orator, will express them by asserting, in a tone of contemptuous irony, that these conclusions, whilst highly satisfactory to the fortunately-placed minority, bring but cold comfort to the majority; that they represent an attempt "to put the clock of progress back," and that the masses of mankind are not very likely to accept them. He will probably go on to say that they are merely a prose rendering of the well-known lines which the sarcastic radical loves—

> God bless the squire and his relations,
> Teach us to know our proper stations;

which last request to the radical seems to be the very height of absurdity; and he will end his attack by appealing to our electioneering instincts, asking us, if we take away the hopes to which at present the masses cling, what new hopes or promises we propose to put in the place of them?

The radical or socialistic theorist, as distinct from the militant politician, will express these same objections in a more logical form, thus: He will remind us that in our analysis of social action we represent the attainment of an exceptional position, and more especially of an exceptional amount of wealth, as the sole motive that can be counted on to induce exceptional men to develop and use their powers. Now this, he will urge, is tantamount to declaring that exceptional wealth is naturally regarded by men as the main condition of happiness; and since it is obvious that exceptional wealth can be possessed by the few only, we are, he will say, convicted of teaching that social progress involves a denial of happi-

The radical theorist will put these same objections more logically. If the desire of exceptional wealth is really the strongest motive, he will say that it follows that most men, since they cannot all be exceptionally rich, must always remain miserable.

ness to the vast majority of those amongst whom social progress takes place ; which, the critic will go on to say, is absurd.

Now even if the conclusions we are discussing did involve in reality all those consequences which would be so depressing to the majority of mankind, yet to prove the conclusions depressing would not be to prove them false ; and few enthusiasts will deny that the object of sociological inquiry is not to reach conclusions which are inspiriting, but to reach conclusions which are true. As a matter of fact, however, the conclusions now in question have by no means that depressing tendency which the radical and the socialist will impute to them.

For, in the first place, none of the arguments contained in the present work have been invoked to prove, or have any tendency to prove, that the many, as distinct from the few, in any progressive country, may not reasonably look forward to a continuous improvement in their condition—to a greater command of the comforts and luxuries of life, together with a lightening or a lessening of the labour necessary to procure them. On the contrary, the majority may look forward to an improvement in their circumstances which it is as impossible for us to imagine distinctly at the present time as it would have been for our grandfathers to imagine the telephone or the phonograph. All that has been urged in this work is as follows : That whatever may be the new advantages which the majority of mankind attain, they will attain them

Now the first answer to this is that the fact that all men will never be equally wealthy does not prevent the conditions of all men from improving absolutely.

23

not by any development in their own productive powers, but solely by the talents and activity of an exceptionally gifted minority, who will enable the ordinary man to earn more whilst labouring for fewer hours, because they will, by directing his labour to more and more advantage, secure from equal labour an ever-increasing product. The conclusion, therefore, is not that the majority in any progressive community may not look forward to indefinitely better conditions, but merely that their condition will not depend on themselves, and that, though the conditions of all may be bettered, they will never be even approximately equal.

What, then, of the argument that, however conditions may be bettered, yet if exceptional conditions are still objects of exceptional desire, the want of these objects of desire will cause a sense of privation amongst the majority?

To this really important question there are two answers.

Another answer is that if inequality in the possession of the most coveted prizes of life implies misery amongst the majority, this evil would be intensified rather than mitigated by socialists, who would substitute unequal honour for unequal wealth.
The first is, that the conclusion now before us —the conclusion that certain of the most coveted prizes of life will always be for the few only—is, whatever may be its consequences, true ; and that its truth is nowhere more clearly evidenced than in the ideal State, as presented to us by the extremest socialists. For we shall find that whatever in the way of equalised incomes these statesmen of cloud-land promise to their imaginary citizens, they do not even suggest that the most coveted social prizes shall be distributed more equally than they are at

the present moment. They, as has been said
already, though they consider themselves the apostles
of equality, recognise that the prosperity, and,
above all, the wealth of the community, will depend
on their securing the very ablest of their citizens
as members of the bureaucracy by whom all
labour will be directed; and they recognise that
these able men, like the present race of employers,
will not develop their ability without some special
inducement. They accordingly propose to reward
them, not by allowing them to retain any exceptional portion of the wealth which they are
instrumental in producing, but by investing them
with exceptional honour; and the desire for such
honour, say the socialists, as a motive to exceptional
effort, "*will be incalculably more efficacious*" than
the desire for wealth. Now if those who make this
assertion attribute to it any serious meaning, they
must mean that men like honour much better than
they like wealth—that they covet it more keenly,
that they will struggle more desperately to win it,
and are more exasperated at not possessing it. If,
however, great wealth is possible for the few only,
and if the majority of mankind are for ever destined
to be without it, such, with regard to honour, is the
case even more evidently. For honour is more essentially confined to the few than wealth is. We can, at
all events, conceive a community composed wholly
of millionaires, supported in luxury by battalions of
labouring automata; but it is impossible to conceive
a community wholly composed of men on whom

honour is conferred as the choicest prize of life, and all of whom—the exceptional and the ordinary —enjoy it to the same degree. The essence of honour is distinction or differentiation ; and it forms a motive for the exceptional actions of the few only because it is withheld from the many whose action is not exceptional. Either, then, in the socialistic State the honour that is to form the reward of exceptionally able men will fail to stimulate their abilities and attract them into the ranks of the bureaucracy because it is not of itself so keenly desired as wealth is ; or if, as the socialists say, it is desired even more keenly, and if it consequently does stimulate exceptional men to struggle for it, the socialistic bureaucracy, with its honours, will excite amongst the mass of the citizens incalculably more envy than the rich excite amongst the poor ; and the millions of average men will be rendered by the want of honour incalculably more miserable than they could be by want of wealth. If, therefore, inequality in the possession of external goods, for which many men struggle, and which only a minority can secure, necessarily means unhappiness for the larger part of the community, this evil at all events is not due to the existing structure of society, but is, on the contrary, so rooted in the constitution of human nature, that even the wildest and completest schemes of social reform are unable to offer us so much as a mitigation of it.

The second answer to the objection, however, is of quite a different, and of a far more reassuring character. It is that the entire supposition on which the objection rests is untrue. The external prizes of life, of which exceptional wealth is the type, though struggled for by many with every faculty they possess, though valued by those who achieve them, and though recognised by men in general as something of which everybody would choose to be the possessor if he could be, do nevertheless amongst average human beings not cause any unhappiness by their absence at all corresponding to the satisfaction which they cause notoriously by their presence. Such an assertion will to many people probably seem self-contradictory. But if it does so, this will simply be owing to the fact that the whole science of the subjective conditions of happiness has been utterly neglected by sociological writers hitherto. The assertion here made, however paradoxical it may sound, embodies one of the most important truths which can claim the sociologist's attention ; and though it cannot be called self-evident, every student of social science should be familiar with it. It forms, indeed, the *pons asinorum* of all social psychology. A brief elucidation of it will be enough for our present purpose.

The final answer is that the unequal distribution of wealth has no natural tendency to cause unhappiness ;

There is a certain minimum of external goods, the desire for which has a physiological basis, and causes, when unsatisfied, misery, disease, or death. Chief amongst such goods are food and, in most climates, clothes and shelter. So far as this minimum is concerned, the desires of all are practically equal ; and they are equal because they arise out of that physical

for men's desires vary. There is equality of desire for the necessaries of life only ; for this desire rests on men's physical natures, which are similar ;

constitution which we cannot alter, and in respect of which we are all similar. But for external goods that are beyond this minimum men's desires vary indefinitely ; and they vary because they depend on the action of the imagination and the intellect, which varies in different men, and in the same men under different circumstances.

In civilised countries the minimum of goods desired is practically not limited to the bare necessaries of existence, and it is difficult to define it with anything like absolute precision. But without any formal definition of it, it is at all events sufficiently distinct to enable us to place in contrast with it those obviously unnecessary goods which make up wealth and luxury. Now luxury is very commonly supposed, in contradiction to what has just been asserted, to represent materialism in its most exaggerated form, and thus to offer a contrast to competence or modest comfort. And it does, no doubt, rest on a material basis ; but competence and modest comfort do so likewise. An arm-chair which costs perhaps thirty shillings is as material as one which, on account of its artistic workmanship, costs four or five times that number of pounds. But so far as wealth and luxury transcend comfort and competence, and possess those peculiar qualities which are held to render them enviable, what they appeal to, and what they are measured by, is not their effect upon the senses, but their appeal to the imagination and the mind. We can easily see this by considering very simple examples, which will show us that the

same external things are luxuries or not luxuries
according to the way in which the mind regards
them. Thus a man will be called luxurious if his
house is of palatial proportions, if he lives under lofty the luxury, for
instance, of a
large house,
ceilings and treads upon shining floors. But the
luxury which the owner finds in existing amongst
these surroundings consists not in any physical
effect which they produce upon his senses as he
moves amongst them, but in a great variety of
complicated relations which exist between them and
his own life, past and future, and of which the senses
take no account at all. Were this not so the poorest
and most destitute might daily enjoy a luxury
superior to that of the millionaire by strolling
through the halls and corridors of our great public
institutions, of which many are far finer than the
most magnificent private houses. A man, again,
will be thought, and will think himself, luxurious if
he travels from Paris to Monte Carlo in a sleeping
compartment with sheets and pillows ; and passengers who have ordinary places, if they are sensitive
to social contrasts, will glare through the windows
enviously at the occupant of this paradise, who has
probably had to pay a hundred francs to enter it.
But let us only imagine that the sleeping compartment
is taken off its wheels and is permanently planted by
the side of some street or road. It will then form
a bedroom which the owner of the pettiest villa
would hardly venture to assign to a maid-of-all-
work ; whilst if three workmen had to sleep in it
instead of three first-class passengers, the agitator

would point to it as an example of the horrors of
overcrowding.   When, therefore, the sleeping compartment is admitted—as it is admitted—to be a
luxury, it is admitted to be so because it is regarded
in relation to a variety of circumstances to which the
senses are quite blind, and which are realised by acts
of the mind and the imagination only.   And with all
wealth and luxury the case is just the same.   Like
comfort and competence, they have material things
for their foundation ; and the material foundation
that supports them is no doubt necessarily larger.
But what renders them more desirable is not the
additional material in itself, but the qualities with
which it is invested by the subtle craftsmanship of
the mind.

Consequently
the desire for
luxury and
wealth, like
the pleasure
they give,
depends on
peculiar mental
powers or
peculiar
mental states.   Just, then, as wealth and luxury depend on the
intellect and the imagination for the larger part of
the pleasure which they give to those who possess
them, so does the desire for them amongst men in
general depend on the action of the intellect and the
imagination also.   Hence, though a desire for wealth
is popularly supposed to be universal, and in a certain
sense is so, it is a desire the non-satisfaction of which
causes a sense of privation only when the imagination and the intellect work in an exceptional way.   Let
us take, for example, some community on the out-
skirts of civilisation which continues to maintain
itself in rude plenty and comfort, but to which wealth
and luxury are merely remote ideas.   If a stranger
suddenly came within its borders carrying a bag
which had in it a hundred thousand pounds, and if

he placed this bag on the summit of a neighbouring mountain and promised to give it to the first man who should get hold of it, every member of this simple community who was not lame or bed-ridden would start for the mountain as fast as his legs could carry him, and the slopes would soon be the scene of a mad and breathless scramble. But if no such stranger came bringing the image of wealth close to them, or if instead of placing his bag on the summit of a neighbouring mountain he showed it to them through a telescope hung up in the moon, not a single heart amongst them would beat quicker at the thought of it or suffer a single pang from the knowledge that it was unattainable.

The reason of this is as follows : Amongst the great masses of mankind the desire for wealth is a speculative desire only. They give, if we may borrow an expression from Cardinal Newman, only a "*notional assent*" to the fact that it is desirable. Wealth means for them no special pleasure which they have experienced, or can represent to themselves, and the repetition of which they crave for ; nor does it mean the satisfaction of any importunate wants. It does not mean for them what a shilling would mean for a starving man. For him the shilling would mean the food for which his stomach clamoured ; and he would feel the want of it as keenly as he would value its possession. So, too, a poor youth separated from his family may crave for a five-pound note, and be miserable at not possessing it, because this will represent the possi-

bility of spending Christmas with them. But no
ordinary man, unless he has lived amongst the very
rich, and his entire view of life has been practically
identified with theirs, has any similar craving for a
hundred thousand pounds, or for a million ; for he
has no personal experience and no detailed knowledge of the peculiar conditions of life which require
such sums to purchase them. Wealth is to him
little more than a name for a power which would
secure for him, if he possessed it, an indefinite
number of indefinite things, if he wanted them ; but
he is under ordinary circumstances no more troubled
by its absence than' he is by the fact that he has
not a fairy for his godmother, or that he does not
happen to be the owner of Aladdin's lamp.

How, then, does it come to be the object of that
keen hunger which is the strongest motive to
activity amongst the men who are the chief producers of it ? What are the exceptional circumstances which convert it from a remote something,
held in a passionless and speculative way to be
desirable, into a near something, craved for, and
eagerly struggled for with the painful industry of a
lifetime ?

The speculative desire for wealth, common to all
human beings, is converted into this practical craving by two causes, which act and re-act upon each
other. One of them is an exceptionally powerful
imagination ; the other is the belief on the part of
any given individual that wealth is a thing which
he actually may acquire if he will only make certain

The desire
ceases to be
speculative
and becomes a
practical crav-
ing only when
the imagination
is exceptionally
strong, and a
strong belief
is present that
the attainment
of wealth is
possible.

efforts, of which he believes himself to be capable. In cases where the necessary efforts are recognised as long and arduous, and the coveted reward as being consequently far distant, the belief of the individual that it is really possible for him to attain it will require the aid of an exceptionally powerful imagination to rouse it into activity, and to keep it alive when roused. In cases where the necessary efforts are obviously extremely slight, and the individual believes that wealth is almost in his hands already, the belief will stimulate his imagination, however feeble it may be naturally, instead of requiring that his imagination should sustain or stimulate *it*. Thus the attainment of wealth being under ordinary circumstances difficult, and requiring intense, anxious, and prolonged effort, a keen desire for it is not ordinarily felt except by men whose strength of imagination amounts almost to genius, and in whom a belief, whether true or false, is developed, that they are capable of creating for themselves this prize which they see so clearly. Warren Hastings, for instance, if his imagination had not been exceptional, would never have had that vision of the past glories of his family which made the desire of restoring them the main motive of his career ; and again, on the other hand, if some sudden and exceptional circumstance, such as the advent of an imaginary stranger with his bag and his hundred thousand pounds, should present every member of a community with a chance of acquiring wealth instantly, the feeblest imaginations would be stimulated to such a degree, that all would find themselves craving for the possible prize equally.

In converting, then, a mere notional assent to the proposition that wealth is desirable into an actual hunger for it, which is painful if not satisfied, the essential cause is a belief that the desired wealth is attainable; and the intensity of the hunger is in proportion to the vitality of the belief. This important psychological truth is very easily demonstrable by a kind of experience sufficiently familiar to most people. If a man who has perfect taste, and a few thousands a year, is buying furniture for his house, and is anxious that every room shall be as beautiful as it is in his power to make it, we all of us know with what eagerness day after day he will stare into the windows of the dealers in old furniture and *bric-à-brac*, and how quickly he will take note of any object that his taste approves. Now if such a man, having admired a cabinet or a piece of tapestry, finds that the price of it is a hundred or a hundred and fifty pounds, he will feel perhaps that it is a little beyond his means; but he will dream of it, long for it, and will never know a moment's peace till he has so arranged his expenditure as to enable him to complete the purchase. But if the price of the cabinet or the tapestry, instead of being a hundred or a hundred and fifty pounds, had been a thousand or fifteen hundred, he would have recognised that the objects were totally beyond his reach, and though they still excited admiration in him, they would

excite no desire.   Here is the great difference
between the necessaries of life and the luxuries.
Men crave for the former, whether they are able to
procure them or no.   They crave for the latter only
in proportion as they feel them to be procurable.
A starving boy does not want a bun the less because
he has not a penny to buy it with.   A man of taste,
with only a hundred pounds to spend, does not
crave for a piece of tapestry at all, if he knows that
the lowest price for it would be not less than a
thousand.

Now under normal conditions the belief that
exceptional wealth is procurable by them is confined
to men with exceptionally vivid imaginations and
with certain exceptional talents and energies that This belief is
correspond to them.   They crave for wealth, in naturally confined to men
fact, because they believe themselves capable of with exceptional imagin-
creating it, and their craving keeps pace with their ations and
exceptional
belief in the range of their capabilities.   The more productive
powers.
wealth they can create, the more they desire to
create.   Their desire for wealth, in fact, unlike
their desire for necessaries, is proportionate not to
their natural wants, but to the extent of their
natural powers.   It follows what may be called *the
law of expanding desire*.   Here, then, is the explanation of the fact which is at first sight so
paradoxical — that whilst the desire of wealth is
the strongest of all motives amongst a minority, the
absence of wealth is not felt as any privation by the
majority ; and so long as the normal conditions that
have just been indicated prevail, and the men who

can really produce exceptional wealth are the only men who believe it to be a thing attainable by them, and are consequently the only men who feel any actual craving for it, all goes well and healthily, and the desire of all classes may be at least approximately satisfied.   Unfortunately, however, the belief that wealth is attainable, though it is naturally confined to men who have exceptional powers of creating it, is capable of being implanted under certain circumstances artificially in men who possess no exceptional powers at all.

It only becomes general by the popularising of false theories which represent wealth as attainable by all, without exceptional talent or exceptional exertion.

A familiar case like the following will show how this is effected.   A man, we will say, occupies an ornamental cottage, which is beautiful in itself, is embowered in beautiful gardens, and also commands views of a picturesque and magnificent park, into the glades of which one of the gates of his garden opens, and which the owner allows him to use precisely as if it were his own.   All his friends tell him, and tell him truly, that there is no such place of its size within fifty miles of London. They envy him his dainty drawing - room, his verandah festooned with roses, his prospect of the timbered park, and his free access to its solitudes. His friends envy him, and he feels himself that he is enviable.   One morning, however, he receives a lawyer's letter, which gives him to understand that he is really the legal owner, not of his cottage only, but of the park and property adjoining, and that with adequate legal assistance he could certainly substantiate his claim to them.   In an instant his whole

It is roused, for instance, in a man who suddenly is told that he has a legal right to an estate which previously he never thought of coveting.

temper of mind with regard to his surroundings is changed. His pride in his cottage is gone, and its place is taken by indignation at having been kept out of possession of the park, and by a feverish craving to acquire it. He goes to law. The case is long and difficult. He lives for months distracted by fear and hope ; and when the case is finally given against him, he comes back to his cottage with his mind unhinged by the shock, contemptuous of the dwelling which once was a source of pride to him, and cursing the prospects which once were his daily pleasure.

Now this craving for wealth, by which the man's life is blighted, has been produced, precisely as such a craving normally is, by the belief on his part that certain wealth is attainable ; but the belief here does not rest on a consciousness that he is able by his own abilities to create or earn it for himself ; it rests on his intellectual assent to a delusive proposition that he has a legal right to it, or, in other words, that the law will make him the possessor of it without any exceptional productive effort of his own. And here we have a counterpart to the socialistic teaching of to-day. It excites, or aims at exciting, an artificial craving for wealth in men who would not naturally trouble their heads about it, by teaching them that they have a right to it, which is wholly independent of any exceptional productive power in themselves, or in any ancestors from whom they might claim to inherit. The only difference between men who are thus deluded, and

Book IV
Chapter 4

the claimant to the park and estate whose case we have been just imagining, is that whilst the latter is deceived into expecting that he individually can be made rich by a law-suit, the latter are deceived into expecting that they all can be made rich by legislation.

The practical craving for wealth is naturally confined to those who have some talent for creating it, and the pain caused by its absence is naturally confined to such men.

The desire for wealth, as something distinct from competence, is a desire which normally affects men only in proportion as they believe themselves to be possessed of power by which they may individually earn it; and so long as men recognise the truth that, apart from rare chances, the powers that earn wealth are the exceptional powers that create it, the craving for wealth which makes the non-possession of it a pain is confined to a minority composed of exceptionally constituted individuals. The absence of wealth amongst the majority causes unhappiness only when false theories with regard to its attainability and men's natural rights to it have produced in the average man an artificial and diseased sensitiveness. There is no surer means of exaggerating inequalities in happiness than the false and pestilent teachings which encourage equality of expectations.

The socialistic theories merely cause a barren and artificial discontent,

And not only do these teachings, so far as they have any effect at all, create private unhappiness and multiply private disappointments, but they give rise amongst masses of men to an impracticable temper, which is the source of many of the difficulties confronting us in the domain of politics, and most of those confronting us in the domain of industry.

The crude and childish philosophy which socialists and so-called labour-leaders endeavour to diffuse amongst the great masses of the population rests, so far as the masses of the population understand it, on the theory that society is composed of "approximately equal units," and that whatever is produced within a community is produced by that community as a whole. Hence the members argue, and the socialists distinctly tell them, that property and capital are merely accidental possessions, which give to those who possess them a purely adventitious power. These teachers add that such possessions, in abstract justice, should be taken from their present possessors and divided amongst the community at large ; and from this it follows that all claims to the profits of capital, as put forward by its present possessors, are, in an abstract sense, unjust. The consequence is that the employed, when stimulated into conflict with the employers, enter on the conflict in a temper which forbids them to be satisfied with any immediate result of it, however favourable to themselves. Whatever advance in wages, or reduction in hours, the employers may have conceded, the employed—so far as they are influenced by the socialistic fallacies of the day—consider themselves still wronged almost as much ·as ever, so long as the employers continue to exist at all ; and thus any cordial understanding between the two classes is made impossible. When the employed strike or agitate for higher wages, they may be compared to a man who maintains that his tailor's bill is

which interferes with that harmonious progress on which the welfare of the many depends.

24

exorbitant, and desires to have a certain portion of the total deducted.   Now if the tailor is reasonable and agrees to take off something, the matter may be easily adjusted to the satisfaction of both parties ; for though the customer may think that the tailor has claimed too much, he admits that to a certain sum the tailor has an undoubted right.   But if the customer were a madman, who believed when he ordered his clothes that in abstract justice he ought to be charged nothing for them, and that any claim on the tailor's part was in reality robbery and oppression, whatever deduction the tailor might consent to make, the customer's grievance against him would remain the same as ever.   It is possible for customers and tradesmen to come to some satisfactory understanding, so long as the demand of the former is that their bills shall not be too high. No satisfactory understanding could be arrived at between them possibly—there would be nothing but friction, constant dunning, and writs—were it known that the customers entertained and meant to act on

the theory that they ought not, in abstract justice, to pay their bills at all.   Now such is the labour-leaders' theory with regard to the employing classes. For a time some part of their bills must unfortunately be paid—that is, some part of their profits be allowed them.   But to these profits they have no real right, and the employed must never be contented until they have absorbed the whole of them. So long as such a theory prevails, no satisfactory progress in the condition of labour is possible, partly because the employed, whatever advantages they may gain, will be no nearer to content than they were before, partly because the employers are constantly forced into a position of unwilling antagonism to men whom they would wish to befriend.

The object of this present work, so far as the question of wealth and its distribution is concerned, has been to show how absolutely false to fact are the theories to which this impracticable discontent is due, and how intellectually ludicrous is the position of the school of thinkers who imagine that such theories represent accurate science. These thinkers, in their dealings with property and capital, in spite of the esoteric admissions of a certain number of them to the contrary, touch the truth in their more popular utterances, only by the process of inverting it, or of putting the cart before the horse. They represent the employing classes as possessing exceptional strength merely because they are accidentally the possessors of capital. The actual truth is that these classes are possessors of capital because they themselves or their fathers have possessed exceptional strength. The arrows of Ulysses were more formidable than those of the suitors because Ulysses shot with a stronger bow than they; but he shot with a stronger bow for the very simple reason that he was strong enough to bend it and they were not. The employing classes contribute to the processes of production not less than the employed; in certain senses they contribute incalculably more, and in every sense they contribute as truly; and they contribute not primarily because they possess capital, but because as a class they possess exceptional faculties, of which the capital possessed by them is at once the creation and the instrument. In other words, the inequalities which socialists regard as accidental are the natural result of the inequalities of human nature, and constitute also the sole social conditions under which men's unequal faculties can co-operate towards a common end.

and to show
that the many
are not a self-
existent power,
Socialists contend that the source of all power is in the multitude. It is impossible to imagine a greater or more abject error. The multitude, or the mass of average men—the men undistinguished by any exceptional faculties—are the source of certain powers, or rather they possess certain powers. That is true; but what may these powers be? Their most striking characteristic is their limitation. In the domain of industry the many, if left to themselves, could produce only a very small amount, which would have, moreover, no appreciable tendency to increase. In the domain of government they could initiate the simplest movements only, and carry out only the simplest measures. The powers which they actually possess under existing circumstances are as much greater than these as the man is greater than the child; but these added powers acquired by the average men, or by the many, do not depend upon average men alone. They are developed only with the development of another set of powers altogether—the powers belonging to the exceptional men or to

the few; and if these latter powers were impaired, the
former would be impaired also. In the domain of
production and the domain of government alike, not
all, but nearly all, the powers of a democracy pre-
suppose the powers of a *de facto* aristocracy, and
although they modify them, they depend upon them. but depend for
Here are the two factors or forces which we can all the powers
never get rid of unless we get rid of civilisation the co-opera-
altogether—the force represented by the mass of
ordinary men, and the force represented by those
who in various ways are more than ordinary. Let
us destroy society a hundred times over, and attempt
to reconstruct it in what way we will, these two
forces will inevitably reassert themselves, and reveal
their existence in the form which society takes, as
surely as a man's figure will give its shape to whatever kind of cloak we hang on it. These two forces
at the present time attract our attention principally
by their activity in the domain of industry, where
they show themselves under the forms of employer
and employed. In order that any satisfactory solution of our industrial difficulties may be arrived at
it is necessary that employers and employed alike
should each recognise the importance of the part
played by the other, the nature and extent of the
other's strength, and the permanent need each has
of the other's strenuous co-operation. It is hardly
to be expected that between these two, serious disputes and difficulties will ever completely cease. In
the interest of social progress it is not necessary
that they should. What is necessary is that what-

ever disputes between these two parties may arise, and however unreasonable or excessive on any given occasion the claims of the few may seem to the many, or the claims of the many to the few, neither party shall regard the other as its opponent, excepting with reference to the particular points at issue ; that the few shall not deal with the many as though the many, in asserting themselves, were rebels, nor the many attack the few, as though the powers of the few were usurpations. What is necessary is that each should recognise its own position and its own functions, and the position and the functions of the other, as being, in a general sense, all equally unalterable, and although admitting of indefinitely improved adjustment, not admitting of any fundamental change.

And what is true of the social forces that are involved in the production of wealth, is true of those that are involved in political government. In political government, just as in the production of wealth, the power of the few has a root in the nature of things as indestructible as has that of the many ; and though the few can produce progress only when the many can co-operate with them, it is not from the many that their power is primarily derived. In the domain of speculative knowledge this is self-evident. The ordinary brains are pensioners of the few brains that are superior to them ; and yet the superior brains are powerless to produce social results, except in so far as the ordinary brains respond to what their superiors

teach them. So it is in economic production, so it
is in political government. The power of democracy is not only an actual power; it is a power from which no society can ever wholly escape; but never —not even when nominally it reaches its extreme development—is it, or can it be, or does it ever tend to be, a power which is self-existent. It always implies and rests upon the corresponding power of the few, as one half of an arch implies and rests upon the other. The whole object of the democratic formulas popular to-day is to deny or to obscure this fundamental truth; and no greater obstacle to general progress exists than the prevalence of the spirit which the acceptance of these formulas en- genders. If there is anything sacred in the rights of the poorest wage-earners, there is something equally sacred in those of the greatest millionaires; and if the latter are capable of abusing their power, so also are the former; but nothing will tend to prevent their abuse of it so much as the recognition that such an abuse on either side is possible. If there is any wisdom and power in the cumulative opinions of ordinary men, there is another kind of wisdom and another kind of power in the ideas, the insight, the imagination, and strength of will which belong to exceptional men; and these last, though they may give effect to what the many wish, do so only be- cause they represent what the many do not possess. What is required to bring our political philosophy— and not only our political philosophy but our political temper—into correspondence with facts is not to

deny the power that has been claimed during this century for the many, but to recognise that this power does not stand alone, and that those other powers represented by the wealthy few are not only essential to the wealth of the few themselves, but also to the prosperity, and most emphatically to the progress, of all.

The progress of all, instead of being incompatible with the fact that the positions of all have no tendency to become equal, assumes, on the contrary, a more and more practicable aspect in proportion to the accuracy with which this fact is recognised ; and that such is the case shall, in conclusion, be briefly shown by reference to the theory of progress which at present deceives the socialists. This theory, which was formulated by Karl Marx, bases itself on the fact, which is indubitable, that the industrial systems of the civilised races of the world have undergone great changes in the past, and may therefore be expected to undergo changes as great in the future. The three most marked stages in the sequence of change referred to are slavery, feudalism, and capitalism ; and the practical conclusion drawn from them by the socialists is that as feudalism arose out of slavery, and capitalism arose out of feudalism, so will socialism arise out of capitalism. This argument is merely another example of those self-confusions by which the socialists are distinguished as reasoners. It is an argument which depends for its whole apparent point on the defective manner in which these various systems—socialism included—

The recognition of the fact that the relations and positions of classes can never be fundamentally altered

(especially when we consider the facts of history to which Karl Marx drew attention)

have been analysed. For, though slavery, feudalism,
and capitalism differ from one another in many most important points, they happen not to differ at all as regards that one particular point in respect of which socialism will have to differ from all three of them. That is to say, in whatever way these three systems differ from one another, they all agree with one another in being systems under which the few, the strongest, the most intellectual, the most energetic, not only controlled the actions of the average many, but received for their exceptional action a correspondingly exceptional recompense. The few who occupied this commanding position differed, at different times, in the nature of the powers which gave them the command. Sometimes it was the great fighters who were paramount, sometimes the great legislators, sometimes the great industrialists. But into whatever mould human society has been cast, with whatever circumstances it has been surrounded, and whatever kind of talent or strength has been most essential to it at given periods, the few who have possessed this kind of talent and strength to the highest degree have, as a whole, and with them their families, invariably occupied a position of exceptional wealth and power. We may deplore this fact or no, but the fact still remains, and consequently the argument of the socialists from the facts of social evolution, when reduced to its true terms, merely amounts to this—that because many social changes have taken place already, but one particular change in spite of these has never taken

place, yet this particular change which has refused to take place in the past is perfectly certain to take place in the future.

The historical evolution of society, however, and the social changes that have taken place, do indeed convey to us a very important moral ; but this moral which the changes convey to us is curiously different from that which the socialists draw from them. They draw from them the moral that because social arrangements have been greatly changed, therefore they can be fundamentally changed. The true moral is that, although they may be changed greatly, they can never be changed fundamentally ; and from this there follows another as its yet more important corollary — that although social arrangements can never be changed fundamentally, they can, nevertheless, be progressively and indefinitely improved, but that real reforms can be accomplished only by those who abandon altogether every dream of fundamental revolution. Many reforms which socialists eagerly recommend, and many wishes which socialists entertain, may meet with the approval and sympathy of the most determined conservatives ; but the error of the socialists is sufficiently indicated by the fact, already remarked upon in the course of this work, that the changes which they advocate, and whose advent they delight to prophesy, leave the possible and approach the absolutely impossible, in precise proportion as these visionaries set value upon them. . Nowhere is the impossibility of such changes more clearly indicated than in the phrases now most

frequently used to indicate their specific nature—
such phrases as "*the emancipation*" and "*the
economic freedom*" of the labourer. These phrases,
if they have any meaning at all, can mean one thing
only — the emancipation of the average man,
endowed with average capacities, from the control,
from the guidance, or, in other words, from the help,
of any man or men whose capacities are above the
average—whose speculative abilities are exceptionally keen, whose inventive abilities are exceptionally
great, whose judgments are exceptionally sound,
and whose powers of will, enterprise, and initiative
are exceptionally strong. That is to say, these
phrases, if they have any meaning at all, mean the
deliberate loss and rejection, by the less efficient
majority of mankind, of any advantage that might
come to it from the powers of the more efficient
minority. "*Economic freedom*," in fact, would mean
economic poverty ; and the "*emancipation*" of the
average man would merely be the emancipation
which a blind man achieves when he breaks away
from his guide. The human race progresses because and when the strongest human powers and
the highest human faculties lead it ; such powers
and faculties are embodied in and monopolised by
a minority of exceptional men ; these men enable
the majority to progress, only on condition that the
majority submit themselves to their control ; and
if all the ruling classes of to-day could be disposed
of in a single massacre, and nobody left but those
who at present call themselves the workers, these

workers would be as helpless as a flock of shepherd-less sheep, until out of themselves a new minority had been evolved, to whose order the majority would have to submit themselves, precisely as they submit themselves to the orders of the ruling classes now, and whose rule, like the rule of all new masters, would be harder, and more arbitrary, and less humane than the rule of the old.

THE END

Printed by R. & R. CLARK, LIMITED, *Edinburgh.*

O

www.ingramcontent.com/pod-product-compliance
Lightning Source LLC
Chambersburg PA
CBHW020236110726
47898CB00004B/1280